PENGUIN BOOKS

SEEING AND BELIEVING

Richard Panek has written about astronomy and science for *The New York Times*, *Natural History*, *Esquire*, and *Outside*. He is the winner of a PEN award for short fiction (which has been broadcast on National Public Radio). He is also the author of *Waterloo Diamonds*, a social history of an Iowa community told through its baseball fortunes. He lives in New York City.

Praise for *Seeing and Believing*

"A terrifically readable history . . . written with the zeal of a born inquisitor."
—*Elle*

"What makes *Seeing and Believing* most unusual is the way Mr. Panek treats the implications of his history, the way he shows how the telescope's development moved in advance of mankind's understanding. To convey this is as vividly as he does requires a skill just as demanding as the ability to explain new developments, namely a capacity to imagine and make clear how people saw things in the past."
—Christopher Lehmann-Haupt, *The New York Times*

"Panek recounts the evolution of the instrument not in stiff, scientific terms, but with the awe and wonder the visionaries he writes about must have felt."
—*Charlestown Post and Courier* (South Carolina)

"A fine diversion for astronomers on those lonely nights when they're waiting for the clouds to clear."
—*San Diego Union-Tribune*

"Panek's history of the revolution wrought by the telescope is a beautiful piece of writing, prose as transparent as a high-grade optic. Reading it will truly open your eyes."
—*Portland Oregonian*

Seeing and Believing

How the Telescope
Opened Our Eyes and Minds
to the Heavens

Richard Panek

PENGUIN BOOKS

PENGUIN BOOKS
Published by the Penguin Group
Penguin Putnam Inc., 375 Hudson Street,
New York, New York 10014, U.S.A.
Penguin Books Ltd, 27 Wrights Lane, London W8 5TZ, England
Penguin Books Australia Ltd, Ringwood, Victoria, Australia
Penguin Books Canada Ltd, 10 Alcorn Avenue,
Toronto, Ontario, Canada M4V 3B2
Penguin Books (N.Z.) Ltd, 182–190 Wairau Road,
Auckland 10, New Zealand

Penguin Books Ltd, Registered Offices:
Harmondsworth, Middlesex, England

First published in the United States of America by Viking Penguin,
a member of Penguin Putnam Inc. 1998
Published in Penguin Books 1999

1 3 5 7 9 10 8 6 4 2

THE LIBRARY OF CONGRESS HAS CATALOGED
THE HARDCOVER EDITION AS FOLLOWS:
Panek, Richard.
Seeing and believing: how the telescope opened our
eyes and minds to the heavens/Richard Panek.
p. cm.
Includes index.
ISBN 0-670-87628-3 (hc.)
ISBN 0 14 02.8061 8 (pbk.)
1. Telescopes—History—Popular works.   2. Astronomy—
History—Popular works.
I. Title.
QB88.P18   1998
522'.2'09—dc21   98–18766

Printed in the United States of America
Set in Times Roman
DESIGNED BY BETTY LEW

Again,

for Meg Wolitzer,

with love

✳

Acknowledgments

The author would like to thank Barbara Grossman for her inspiration, Dawn Drzal for her expert editorial guidance (even while carrying a heavier burden than usual), Henry Dunow for his ongoing faith and friendship, and especially Gabriel and Charlie for indulging their dad again.

Contents

The children have built something out of an orange crate, something preposterous and ascendant. . . .

—John Cheever

Prologue

On January 15, 1996, the universe grew by forty billion galaxies.

That morning a team of astronomers made public an image that reached farther across space and therefore farther back in time than any other photograph ever. This was the Hubble Deep Field. For ten consecutive days the preceding month, the Hubble Space Telescope had aimed into a single speck of space and extracted all the light it could get. The result, as the members of the HDF team explained it, was a kind of geological dig through layer after layer of a "core sample" of the universe. Although the circumference of the hole that Hubble had drilled through the heavens was at the limit of what the naked human eye can distinguish—the equivalent of a dime at a distance of seventy-five feet, or, as one member of the team more evocatively suggested, a grain of sand at arm's length—it seemed to contain at least fifteen hundred to two thousand galaxies. If one traditional definition of infinity was the number of grains of sand in the universe, a new metaphor now presented itself: the number of universes in a grain of sand.

What actually grew that morning, of course, wasn't the size of the universe, but our understanding of it. And the photograph wasn't technically a single image, but a computer-enhanced montage of 276 separate exposures taken during 150 Earth orbits. Even the figure of forty billion, while impressively specific and suggestively scientific, was deceptive. An hour before the press conference, at the annual winter meeting of the American Astronomical Society in San Antonio, Texas, a couple of team members had hunched over a hand calculator, punched some keys, and come up with a number. If fifteen hundred to two thousand galaxies crowded the Hubble Deep Field, and thirty million such near-microscopic fields covered the sky, then fifty billion galaxies altogether (or forty billion more than the previous best guess) wasn't in fact an unreasonable estimate—but it was nonetheless a rough estimate, at best.

Still, it did the job. Even while downloading the HDF, the astronomers had recognized in its gathering pixels the kind of document that could alter the course of astronomy in particular, science in general, and—why not?—society itself. It wouldn't be the first time. Over the previous four centuries significant advances in telescope technology often had led to radically new understandings of the universe, and in the HDF, Hubble had detected objects ten times as faint as what the most powerful ground-based telescope could see, and double Hubble's own previous best. The photons that would become the HDF—indivisible packets of energy radiated by innumerable pieces of matter—had begun their journey maybe ten billion years earlier, traveling some sixty sextillion (that's 60,000,000,000,000,000,000,000) miles before reaching the rims of Hubble's 94.5-inch-diameter primary mirror, which focused them onto a 12.2-inch secondary mirror, which redirected them into a host of scientific instruments, which, after translating them into electronic signals of zeros and ones, beamed them to a tracking and data relay satellite, which ricocheted them to a ground station in White Sands, New Mexico, which, after translating them into radio signals, zapped them back up into the stratosphere and toward a communications satellite, which bounced them earthward again to the Goddard Space Flight Center in Greenbelt, Maryland, which forwarded them via telephone circuitry to the Space Telescope Science Institute on the Johns Hopkins University campus in Baltimore, Maryland, where they took up residence inside computers until an astronomer called them up on a screen, at which point the zeros and ones regathered themselves into swatches of light and dark that, approximately two feet and 1/500,000,000th of a second later, reached the eyes of astronomers, who could hardly believe what they were seeing. "As the images have come up on our screens," the director of the Space Telescope Science Institute declared, "we have not been able to keep from wondering if we might somehow be seeing our own origins in all of this." He dubbed the emerging image "the double helix of galaxy formation"; another astronomer likened it to the Dead Sea Scrolls.

The relationship between the telescope and our understanding of the dimensions of the universe is in many ways the story of modernity. It's the story of how the development of one piece of technology has changed the way we see ourselves and of how the way we see ourselves has changed this piece of technology, each set of changes reinforcing the other over the course of centuries until, in time, we've been able to look back and say with some certainty that the pivotal division between the world we inhabit today and the world of our ancestors was the invention of this instrument. Of course, the same is true for any number of other modern tools, including some with especially resonant associations—movable type, the clock, the air pump, or the telescope's first cousin the microscope—but no other instrument has consistently addressed the question of our place in the universe as directly as the telescope. It's what a telescope *does;* it's what we have designed it, and then refined it, and refined it, and refined it, to *do:* address our place in the universe, literally. To size up all of space and figure out where we are in it. And once we'd glimpsed the telescope's potential, its purpose has never *not* been to seek the boundaries of the universe. Once we'd figured out what the telescope could do, the goal has always been to make it do *more.* In a sense the members of the HDF team were simply following an impulse dating back to the first optic tubes: to look deep into the sky to see . . .

To see what?

They couldn't say. One of the most popular, most enduring misperceptions about science is that it proceeds in an orderly, linear fashion. That science works like this: Hypothesis precedes observation, which proves it right or wrong. That the need to make an observation creates the need to use an instrument.

It does, and it doesn't. Much of the preparation for the Hubble Deep Field did indeed proceed in this kind of cause-and-effect fashion. A year earlier the director of the Space Telescope Science Institute had decided to use much of his annual allotment of discretionary observing time on the telescope—a perk that came with the title—to investigate galaxy formation. A panel of advisers had suggested that in order to meet this goal, the Hubble telescope should take the deepest image of the universe ever, and they'd found a speck of sky that satisfied the necessary criteria—one sufficiently free of stars in our own galaxy, and completely free of known nearby galaxies, that it would allow a clear peephole through space.

But that didn't mean the Hubble team knew what it would find. The truth about science is that it also proceeds in an unruly, intuitive fashion. That sometimes observation precedes hypothesis, and what creates the need to make the observation is a desire to see what a particular instrument can do. That sometimes the best answer a scientist could want is more questions. In the case of the Hubble Deep Field, the questions it raised were twofold: Why did this image contain so much more light than astronomers expected? And why so much more dark?

More light, because the image revealed galaxies in a number and variety without precedent. Not only did a total of fifty billion galaxies in the universe reflect a fivefold increase over previous best estimates, but the galaxies in the Hubble Deep Field came in colors, shapes, and sizes that nobody recognized. In astronomy, distance equals time; the farther away in space an object is, the farther back it is in the past. Astronomers seeing the multitudes of unfamiliar bluish blotches and whitish whorls in the HDF had to wonder if they belonged to galaxies in the process of formation, perhaps as early as one to two billion years after the birth of the universe.

And more dark, because between all these individual sources of light lay the black of space. Usually photographs approaching this depth would contain, as one astronomer put it, "wall to wall galaxies." Instead the wall appeared to have gaps. In part this clarity was due to Hubble's unprecedented resolution, the ability to produce sharply defined images. But astronomers also had to ask themselves if they were looking not only at galaxies in the earliest formative stages but past them, to a time before galaxies, before stars—to a time just plain *before*. And if that was the case, they then had to wonder if they possibly were within striking distance of the ultimate goal that astronomers had been seeking since the invention of the telescope: the ends of the universe.

Telescopes turned. Around the world and up in space, instruments swiveled toward a seemingly barren patch of sky near the handle of the Big Dipper. Astronomers abandoned their observing routines to push the limits of their technology and their imaginations, in large part in an attempt to pinpoint precisely where and what the objects in the Hubble Deep Field were, but also partly for the same reason the HDF team had decided to spend ten precious days excavating a hole in the universe, the same reason any number of observers over the past four centuries had first picked up an instrument that promised a closer look at what's out there: just to see what they could see.

Part I

*

SEEING

✳

CHAPTER ONE

The New World

Tube of lead, two disks of glass: The answer, when at last it arrived, appeared to be simplicity itself. For several years lenses of the right shape had lain within reach of anyone entering a spectacle maker's shop. For three centuries spectacles themselves had been popularizing the magnifying properties of curved glass. For the past couple of millennia it had been common knowledge that if you want to eliminate distracting light and concentrate your powers of observation on a distant object, use a tube. In many ways the arrival of this latest wonder in an age of wonders seemed to have been only a matter of time. Not so the transformation it then underwent. That, nobody could have foreseen, even though the one ingredient essential for the conversion of this potential plaything into an instrument of mathematical and philosophical investigation—for its invention as the telescope, in effect—presumably had been present at the Creation: curiosity about a Creator.

On the fourth night after the new Moon that rose at

the end of November 1609 over the rooftops of Padua, a town outside Venice, a professor of mathematics at the local university, Galileo Galilei, raised the new instrument toward the sky. He would have had no reason to think he was the first to exercise this impulse. For more than a year the instrument had been circulating in one form or another across the Continent. This wasn't even the first time that *he* observed what he witnessed that night: the Moon of old, the Moon of always, only bigger. But he returned to it now, as he often had returned to it, because under magnification the Moon yielded surprises—discrepancies between how it had appeared since the beginning of time and how it actually, apparently, was.

The line dividing the light regions of the thin crescent from the dark, for instance: It wasn't the long, smooth, graceful, perfect arc it otherwise appeared to be. Instead it was broken, serrated, almost as if it were ridged—as if the surface itself varied, as if the landscape of the Moon were interrupted by elevations and depressions. The darkness of the Moon proved equally deceptive. Within its depths Galileo located pinpoints of light. For two hours, until the Moon slipped out of sight, he watched as the pinpoints sprouted and spread, their buds of light widening, emerging out of the black and into the white of the night exactly as if they were great heights catching the morning sun. He had to struggle against the breath that clouded the glass, against the blood that shook the hand that steadied the tube. But the more he searched, the more he found. He made out mountains. He divined valleys. He saw seas.

Galileo might not have been the first to turn a spyglass on the heavens, but he had spent the past several months improving the instrument and his familiarity with what it could do, and as he did so, he had begun to entertain an intriguing interpretation of the heavenly bodies. When he had directed his spyglass at the stars that didn't move in relation to the other stars—the fixed stars—he found that their appearance didn't change under magnification. Yet when he had looked at the stars that did move across the heavens—planets, they were called, after the Greek word for "wanderers"—he found they *did* change. They weren't the pinpoints of light they always had appeared to be, but "globes," as he soon wrote, "perfectly rounded and definitely bounded, like little moons."

Planets looking like the Moon looking like the Earth: What Galileo had been observing wasn't simply impossible to see without the new tube of long seeing. It was impossible, period. It was inconceivable, almost. Yet there it was: a chain of logical associations leading, if not to an inescapable conclusion, at least to an unavoidable question: Were the planets in fact other Earths? More to the point, was Earth another planet—one more mere wanderer in the heavens?

In many respects this line of investigation violated common sense. What was terrestrial was terrestrial, and what was not was celestial, and to say otherwise made a mockery of the evidence of the senses and the wisdom of the ages. The greatest philosophers who ever lived had used this distinction as the basis for their beliefs, and those beliefs formed the foundation of civilization:

its astronomy certainly, but its philosophy too, its religion and its physics, its understanding of the relationship between Heaven and Earth, between God and Man. Equally to the point, at least, this reasoning contradicted the evidence of the senses. Drop a rock, and it doesn't land a few miles away or even a few paces away; it lands at your feet. You didn't have to be an Aristotle to question the idea of a speeding, spinning Earth.

In other ways, however, what Galileo saw made compelling, possibly perfect sense. The preceding two centuries had produced any number of discoveries that ran counter to the wisdom of the ages: lands of which the ancients knew nothing, peoples whose salvation the Bible failed to take into account. These discoveries had expanded the common perception of the world, and with it perception itself, the very concept of common sense. If the world could learn to accommodate the addition of terrestrial lands and their mysteries—ivory and tobacco, Africans and West Indians—then why not celestial lands? If a New World, why not new worlds?

Exactly two hundred years before Galileo first raised his spyglass to the sky, the world—which is to say, that portion of the world near the center of which a scholar in Florence or Constantinople might imagine himself— was innocently living out its final year of isolation. According to the standard maps of the day, how the world of 1409 saw itself was exceedingly simple. Draw an *O,* then make a *T* inside it. The orientation, as that word suggests, is to the east. What lies on top, above the crossbar of the *T,* is Asia; to the bottom left of the vertical support, Europe; to the bottom right, Africa. Simple, and sufficient: With a few notable exceptions—a Marco Polo here, a Crusade there—travelers didn't journey to faraway lands. Why would they? First, there was only so far to go. If more of the world were habitable, the assumption went, then people already would have inhabited it. Africa, for instance, might in fact extend some distance south of the shores of the Mediterranean Sea—the body of water represented on the maps by the upright in the *T*—but to venture toward a place where the Sun was sure to pass overhead was to risk burning wood and cloth, charring hair and flesh. As if to punctuate this proscription, mapmakers traditionally encircled their three admittedly, understandably incomplete continents with the *O* in the so-called T and O map, an impassable body of water that engulfed the world beyond the known borders, the Ocean River.

Second, even if there *were* places to go, there was little point in doing so. The point of earthly endeavors, scholars and clergy agreed, and had agreed for hundreds of years, shouldn't be to investigate a world as impermanent, unstable, and ultimately unrewarding as this, but to prepare oneself for the next. Toward that end mapmakers provided not details of distance and dimension, but art—visual sustenance both secular and sacred, representations of man's place in the world and reminders of the glory of God, from the figures of the final resurrection crowding the corners of the manuscript to the natural prominence of the city of Jerusalem at the point where the *T* crossed itself, in the center of map, Christendom, and universe.

The next year all that changed. In 1410 a Florentine

scholar published a translation of a Greek text that he had found in Constantinople. This was the *Geography* of Claudius Ptolemaeus, or Ptolemy, an astronomer and cartographer who lived and wrote in the second century after the birth of Jesus. The drawings that originally accompanied Ptolemy's text had disappeared, but cartographers of the early fifteenth century found that by following his written descriptions, they could reconstruct these maps. Where their knowledge of Asia ended at the Ganges, Ptolemy described a continent that extended at least fifteen degrees of longitude beyond India and included numerous islands. Where their conception of Africa ended just inside the Mediterranean coastline, Ptolemy's Africa extended at least fifteen degrees of latitude beyond the Equator, deep within the "torrid zone" where mathematicians had calculated the Sun would pass overhead (had any heads been there for it to pass over). Most important for the imaginations of the philosophers and navigators, the traders and geographers who gathered in the streets of European ports to discuss each new piece of knowledge, what Ptolemy had to say about the Ocean River wasn't forbidding. It wasn't welcoming, either. Ptolemy treated the Ocean River as if it were simply there, like any other waterway, one more potentially navigable route to facilitate the trade of riches.

The Portuguese capitalized first, and often. Under the sponsorship of Prince Henry the Navigator, Portuguese fleets reached the Canaries in 1416, Madeira in 1420, the Azores a decade after that. By 1434 Portuguese ships had passed Cape Bojador, which marked the great western bulge of the continent; by 1473, the Equator itself; and then, in 1488, the southern tip of Africa, after which the seas opened onto the vast expanse of the Indian Ocean and the promise of clear sailing to the east—a promise that in 1498 Vasco da Gama fulfilled by reaching India. In so doing, he completed the search for an overseas route between West and East that had taken the better part of a century.

But the world extended in more than one direction. While the Portuguese sought a southern route along the Ocean River, somewhere past the seemingly endless shores of Africa, some navigators wondered if there might be a more direct approach to the spice islands of the East. The world, after all, was round. It didn't feel round, of course, and taking its roundness into account required an act of will. But the fact that the Earth was a globe was readily apparent, and had been since antiquity. Eighteen hundred years earlier Aristotle had noted in his *On the Heavens* that the true shape of the Earth was written in the shadow that crossed the Moon during every lunar eclipse; it was inscribed on the celestial sphere by the new constellations that presented themselves to travelers who ventured to the ends of the Earth, and then beyond. "For this reason," Aristotle wrote, "those who imagine that the region around the Pillars of Hercules"—the Strait of Gibraltar—"joins on to the regions of India, and that in this way the ocean is one, are not, it would seem, suggesting anything incredible."

So seas were navigable, and the world was round. What might this suggest to a sailor looking for an excuse to journey due west from, say, Spain? Again, a new translation of an ancient manuscript provided encouragement. Strabo, a geographer from the time of Christ whose work first appeared in Latin around the middle of the fifteenth century, took Aristotle's hypothesis to its logical conclusion: "The habitable world forms a complete circle, itself meeting itself, so that, if the immensity of the Atlantic Ocean did not prevent, we could sail from Iberia to India along one and the same parallel over the remainder of the circle."

To the north lay Europe; to the south, Africa; to the east, Asia. What lay west was Asia again, eventually. How far west was another matter entirely, and Christopher Columbus, for one, spectacularly miscalculated the distance. In preparing his arguments for a westward expedition, he underestimated, then underestimated further, the circumference of the globe; he overestimated, and overestimated further, the span of Asia; until, by his optimistic calculations, a ship sailing the parallel from Spain would reach landfall right around the point where he, in fact, did. For this reason Columbus assumed he'd reached one of the islands that Ptolemy had described as belonging to the farther reaches of Cathay.

While further calculation and subsequent voyages established that Columbus hadn't reached Cathay, what this land *was* remained open to debate. Even the first circumnavigation of the globe, completed by the survivors of Magellan's crew in 1522, confirmed only the enormous scale of the lands that Columbus had discovered, not their nature. Some argued the new lands were an extension of Asia—a considerable extension and an

undeniable advance on previous geographical knowledge, but still, no challenge to the traditional conception
of a tripartite globe. Others claimed for the mainland
the status of a new landmass altogether, a fourth continent to rival Asia, Africa, or Europe in size, a half world
unto itself. Yet whatever the New World might turn out
to be, it was undeniably *new,* and that alone was sufficient to distinguish it. "New islands, new lands, new
seas, new peoples," wrote the Portuguese Pedro Nunes
in his *Treatise of the Sphere,* in 1537, "and what is
more, a new sky and new stars."

The very idea of newness was a novelty. The word
"novelty" itself was one of several forms of the Latin
verb *innovare*—meaning "to make new again," as in the
rediscovery of ancient manuscripts—that slipped into
the language, in the late 1400s and throughout the
1500s, to meet the demand for descriptions of what was
new: innovation, novella, novel. For a thousand years,
since the fall of Rome and the burning of the library at
Alexandria, the peoples of the Mediterranean had with-
drawn from knowledge and from themselves. Now,
invoking the right of succeeding generations to write
history, scholars recast the preceding thousand years as
the Middle Ages, in the process both distancing themselves from their immediate predecessors and implying
a kinship between beginning (the Golden Age, Greece)
and end (now, us).

For a scholar of the fifteenth century, no calling was
nobler, no aspiration higher, than to travel to Constantinople, unearth an ancient manuscript (it hardly mat-
tered which ancient manuscript), and translate it. In

those days any number of such volumes—Aristotle,
Plato, Eratosthenes, Ptolemy—were reaching the light
of day in new translations, and everywhere the lesson
seemed the same: The ancients had done it first and
done it better. This made perfect sense; these were the
greatest minds who ever lived, and since the store of
knowledge was finite, they had simply mined it first.
But this realization placed a distinct burden on the new
generation, the responsibility to rediscover the past and
to learn the lessons well.

This they did. A cottage industry even developed
among craftsmen who celebrated the cult of newness
and who found an appreciative audience equally eager
to celebrate it. *Nova reperta,* these works of art were
called: "new discoveries." Sometimes they depicted a
wonder of the age, sometimes the wonders of the ages:
from ancient times, the water mill, the astrolabe, olive
oil, the cultivation of sugar, the production of silk; from
medieval times, the windmill, the lodestone, gunpow-
der, stirrups, spectacles, the process of distillation; from
modern times, printing, exploration, oil colors, even
copper engraving, the process employed in the manu-
facture of many of these same *nova reperta.* The implication was clear, if not entirely intentional: The
accomplishments of the present age could hold their
own against anything the past had to offer.

And not just hold their own: Simply knowing that the
world was round hadn't gotten the ancients anywhere.
What had brought modern explorers to new lands was
their ability to take a flat surface upon which three continents crowded inside a circle and imagine west bend-

ing back, wrapping around, and meeting east. In time the high regard in which the moderns held their own accomplishments came to carry a further implication: not just different, not just new, but in some way *better.*

"Our age today," wrote the sixteenth-century French professor of medicine and court physician Jean Fernel, summarizing the sentiments of his scholarly peers, "is doing things of which antiquity did not dream." In 1539 a Paduan philosopher proclaimed, "Do not believe that there exists anything more honorable to our or the preceding age than the invention of the printing press and the discovery of the new world; two things which I always thought could be compared, not only to Antiquity, but to immortality." Gómara, in his *History of the Indies,* was at once more exacting and more exalting in his estimation of what would endure: "The greatest event since the creation of the world (excluding the incarnation and death of Him who created it) is the discovery of the Indies." And the French political theorist Jean Bodin, in an evaluation of his times that came to serve as a rallying cry for his contemporaries, wrote, "The age which they call golden, if it be compared with ours, would seem but iron."

The ancients, it turned out, weren't always right. "With all due respect to the renowned Ptolemy," wrote one Portuguese sea captain, "we found everything the opposite of what he said." Ptolemy might have been correct in his overall conception of broader travel, but he was mistaken in many details—for instance, in holding with ancient tradition and placing a mysterious terra incognita at the southern tip of Africa that, had it been where he said it was, would have blocked that sea route to the East.

Personal observations, it further turned out, might prove even more valuable than the word of ancient authority. "What I have said," Fernández de Oviedo remarked about his writings on the New World, "cannot be learnt in Salamanca, Bologna or Paris." It was one thing for Socrates to muse, "I believe that the earth is very large and that we who dwell between the Pillars of Hercules and the river Phasis"—in the Caucasus—"live in a small part of it about the sea, like ants or frogs about a pond, and that many other people live in many other such regions." It was quite another—and arguably quite a better—method to find those people, to explore those regions. *Ne plus ultra*—"No farther"—read the inscription on the mythical Pillars of Hercules, marking the ancient terminus of navigation and of knowledge. Now, *Plus ultra* proclaimed an Age of Discovery: "Farther yet."

As remarkable as any or all of these physical achievements were, at least equally noteworthy was an intellectual achievement that had to accompany them—a newfound ability to appreciate the accomplishments themselves, to see them not only as different, or new, or even better but as part of a historical process: to put them in perspective.

Perspective—from the Latin for "to look through"—was one more novelty in an age of novelties, but it was in many ways the defining one. For an age of new things to see, the artists of *la nuova arte* had created a new way of seeing. They had taken a two-dimensional surface, applied the principles of geometry, and discovered a way to represent the workings of the three-dimensional world. In their hands there vanished the sacred backdrop of old—the golden or black curtain against which saints and Savior had enacted their symbol-drenched dramas. Instead what stood revealed, receding to infinity, were the details of everyday life: walls, doors, windows, and what you could see through them: people, animals, sky. Yet as brazen as the artistic breakthrough happening in the streets of Florence—the tricks of the eye that Filippo Brunelleschi and his disciples were executing so that, in the words of one awestruck observer, "the spectator felt he saw the actual scene"— was the insight that had to accompany it: that there was a curtain to part.

The uniform backdrop of old was a presence that didn't fully reveal itself until it was absent. Only then did it yield its secrets: how it had represented an unwavering, unthinking assumption; how its uniform surface of gold or black provided a sensibility unto itself; how that sensibility reflected a singular point of view— saints of a certain size, Virgin of another, Jesus of a third, all arrayed according to a hierarchy that could belong only to the mind of God. But now it was gone, and what replaced it was a point of view that could belong only to the human eye—and only to an eye seeing the world as if for the first time.

One innovation perhaps more than any other eased this transition. What artists had been advocating was nothing less than a new way of looking at the world, and spectacles were it. They were a tool to use, yes, but also a metaphor to appreciate, something to hold and something to hold on to. Spectacles actually predated *la nuova arte* by a good century and a half, but in time they served as a common symbol for what artistic perspective could accomplish. An instrument that literally brought familiar settings into focus provided a powerful everyday example of what it might be like figuratively to see the world new.

In the hierarchy of innovations according to *nova reperta* engravings, spectacles would have belonged to the Middle Ages. In fact, knowledge of the magnifying properties of precious stones probably dated to antiquity, but the use of glass expressly to correct eyesight was a phenomenon of the mid-thirteenth century. Specifically these magnifiers—thick sections of glass spheres to place directly atop reading material—addressed the diminishing ability to see near objects clearly that often accompanies advancing age. The glass in the first eye spectacles was biconvex—that is, it bulged outward on both sides—and looked like a lentil, or *lens* in Latin. Biconvex lenses brought near objects into focus, and generally anyone afflicted with farsightedness could walk into a spectacle maker's shop and find reading glasses of the approximate appropriate strength—a focus of twelve to twenty inches.

Not so those with nearsightedness, the diminishing ability to see distant objects clearly. For this problem concave lenses—lenses that curved inward on both sides—were necessary, but not just any concave lenses. Unlike spectacles to correct farsightedness, these didn't fall within a conveniently narrow range of possibilities;

instead spectacles for nearsightedness needed to focus at the specific distance at which the eyesight failed, and the weaker the eyesight, the greater the distance at which they would need to focus.

Sufficiently weak convex, sufficiently strong concave: The pieces were now in place. Around the turn of the seventeenth century almost any spectacle maker's shop was stocking the two kinds of lens that sooner or later someone was going to hold twelve to fourteen inches apart and train on a distant object, and find it close enough to touch. By 1608 someone had.

Late that year spyglasses began appearing everywhere, and all at once. They surfaced at the annual autumn fair in Frankfurt, courtesy of an anonymous vendor, although his sample was cracked, his manner unpleasant, his price unreasonable. On October 2 a spectacle maker named Hans Lipperhey, from Middelburg, in the Flemish province of Zeeland, applied before the States General of The Hague for a patent on a "certain instrument for seeing far." Within two weeks two other applicants had filed claims with the States General as well: Jacob Adriaenszoon (also known as Jacob Metius) of Alkmaar, a city in the province of Holland, who said that for two years he'd been working with lenses in order to understand their magnifying powers and that he'd improved a device to the point "he can see things as far away and as clearly as with the instrument which was recently shown to Your Honors by a citizen and spectacle-maker in Middelburg," and Sacharias Janssen, another spectacle maker in Middelburg, "who says that he also knows the art, and who has demonstrated the same with a similar instrument." If nothing else, these rival claims succeeded in convincing the States General that the invention didn't deserve a patent. As an adviser put it, "the art cannot remain secret at any rate, because after it is known that the art exists, attempts will be made to duplicate it, especially after the shape of the tube has been seen, and from it has been surmised to some extent how to go about finding the art with the use of lenses."

The solution turned out to be simple, once you knew it. It was only one of many combinations available, and it was a most unlikely one at that. Conceivably, constructing one could have involved deciding: among convex lenses of various strengths; among concave lenses of various strengths; which to hold against the eye and which to hold away from the eye; what distance from the eye to hold each; even, to begin with, the number of lenses—one, two, or more. In the end the version that wound up circulating throughout Europe required placing a weak convex away from the eye and a strong concave near the eye—which is to say, using a *weak* magnifier, and then only at some distance, with a lens that actually *shrank* images, and that one, right next to the eye. Still, put the two together in that configuration, and they worked, somehow. And once they were in place, it was a design that was easy to copy. It belonged to no one, and to everyone.

It spread quickly. Spyglasses had reached Paris and London by the following spring, Milan in May, and Venice and Naples by August. *Word* of the spyglass spread even more quickly. In November 1608 a

Venetian theologian named Paolo Sarpi first heard about a device for seeing things at a distance as if they were near. Rumors of miraculous devices were hardly unusual, and often unreliable, but these proved persistent, and the following March he wrote to Jacques Badovere, a friend in Paris, to ask if this rumor from abroad was true. Even so, it wasn't until another couple of months had passed that he mentioned what he'd heard about the spyglass to a close friend who was visiting him in Venice, the professor of mathematics at the University of Padua, Galileo Galilei.

Galileo had to know: *Was* it true? From what Sarpi had heard of the spyglass from Badovere, it was indeed, and Galileo pronounced himself "seized with a desire for the beautiful thing."

Presumably he also was seized with a desire to advance his lot in life. As a professor of mathematics he possessed some knowledge of optics; as the inventor and manufacturer of a navigational instrument called a proportional compass he had an even more extensive knowledge of mechanics. Here was a device he probably could figure out how to construct by himself, and very possibly improve, and, in so doing, secure a reward from the Venetian Senate—but only if he could present his claim before anyone else. In fact, Sarpi had mentioned the spyglass only because a spyglass salesman was reportedly in Padua even now, stopping on his way to Venice.

Galileo returned to his workshop in Padua at once and notified Sarpi to advise the Venetian Senate not to accept any rival claims until he could return and present his own. By the end of the month Galileo had done just that. For several days he led the elders of Venice to the tops of towers to see for themselves the sails on the ships at sea, many miles and several hours beyond the reach of the unaided eye, and as his reward he received a new contract at the university, at a greater salary for life.

Galileo himself never claimed to be the instrument's "first inventor," the proper term at the time for the person who initially constructs a new device. Instead he cast himself as an "inventor" of the instrument, someone who had duplicated a device that was already there. Not that he had simply copied the spyglass; he'd had to figure it out for himself. When he withdrew to his workshop in Padua, he tried various combinations of lenses until one worked: a plano-convex lens (plane on one side, convex on the other) at the far end, a plano-concave lens near to the eye.

But he hadn't stopped there. He'd improved it. First he used his knowledge of optics to figure out the mathematical relationship at the heart of the device's power to magnify: the ratio of the focal lengths. A spyglass with a lens at the far end that brings images to a focus twelve inches away and a lens at the near end that has a focus of four inches will be able to magnify images three times—or twelve divided by four. Once he'd figured out this formula, it would have been relatively simple for someone with Galileo's technical expertise to grind new lenses to exploit this mathematical relationship to best advantage.

The instrument he presented to the doge seemingly

brought objects eight or nine times closer and made them more than sixty times larger—a tremendous improvement over rival spyglasses, which magnified objects as if they were three times closer and nine times larger. When Galileo later published accounts of his introduction to the new device, he gave credit where it was due—"a rumor came to our ears that a spyglass had been made by a certain Dutchman"—but he also drew a distinction. The first inventor of the spyglass had arrived at the proper construction presumably by "casually handling lenses of various sorts"—that is, by luck. He, Galileo, incited by the news, "discovered the same thing by means of reasoning." His "reasoning," in fact, bore a close resemblance to trial and error, but still: He not only managed to duplicate the model that already existed; he began improving it significantly, and this, to his mind, was reason enough to reward him. (And reward him properly. When Galileo decided that a raise and tenure in Venice weren't adequate compensation for his achievement, he struck a better bargain with the court in his native Florence.)

Yet what truly distinguished Galileo's initial presentation of the instrument was that he saw its potential—not that a magnification device would be handy for investigating the heavens, but that a humble novelty item could deliver on an age-old promise of *nova reperta* dreamers. For hundreds of years rumors had attended every advance in knowledge about the wondrous properties of glass in general and spectacles in particular. Some reputable reports had even suggested a familiarity with instruments that produced magnifications as high as a thousand times. "Thus from an incredible distance," Friar Roger Bacon had written of the "wonders of refracted vision" in his *Opus majus* of 1267, "we might read the smallest letters and number grains of dust and sand owing to the magnitude of the angle under which we viewed them." In the 1570s Thomas Digges wrote that his father, Leonard Digges, "hath by proportionall Glasses duely situate in convenient angles, not onely discovered things farre off, read letters, numbred peeces of money with the very coyne and superscription thereof, cast by some of his freends of purpose uppon Downes in open fields, but also seven myles of declared wat hath beene doon at that instante in private places." And William Bourne wrote in his *Inventions or Devices* of 1578, "For to see any small thing a great distance of from you, it requireth the ayde of two glasses, and one glasse must be made of purpose, and it may be made in such sort, that you may see a small thing a great distance of, as this, to reade a letter that is set open neare a quarter of a myle from you, and also to see a man four or five miles from you, or to view a Towne or Castell, or to see any window or such like thing six or seaven myles from you."

Such instruments, however, didn't yet exist. Despite their seeming certitude, these writings (and many others) were speculations or embellishments. Sometimes their authors were fantasizing about what might happen with the right combination and quality of lenses. Sometimes they were repeating or enhancing already exaggerated claims for existing instruments. Mostly, though, they were indulging in natural magic, the province of magi who studied the ways that nature can deceive the senses. Still, if nothing else, the enduring interest in the subject reflected a growing fascination with the possibility, if not the certainty, that glasses of great magnification were a miracle waiting to happen. A lens here, a lens there, and presto!—distant images close enough to touch.

Anyone living in anticipation of that miracle, however, easily might have missed it. In its original form the spyglass produced results that were far more modest than its prophets had foreseen. Giambattista della Porta, one of the foremost magi of the era and an expert on optical magnification, wrote in a letter dated August 28, 1609, in response to a query regarding the reports of Galileo's great success with the elders of Venice in the previous week: "About the secret of the spectacles, I have seen it, and it is a hoax." He went on to describe the placement of the lenses and even supplied a drawing, the first-known illustration of a spyglass. But the device, he said, was similar to something he had proposed twenty years earlier—and it was, roughly. In the 1589 edition of his widely popular *Magia naturalis,* he had described the properties of "Concaves" and "Convexes" accurately, and it was even likely that he'd constructed magnification instruments, perhaps distributing them to friends as eyesight correctives. In the end, though, he wound up characterizing his findings as useful for "merry sports." He failed to see what a modest magnification of two times, or even six times, might have to do with the magnifications in the hundreds or thousands that he and his fellow magi had been anticipating. Like the mountains on the Moon, the instrument very well might have been there before anyone realized it, and even when a Galileo came along and claimed that here at last was the miracle that centuries of seers had predicted, a certain restraint was understandable and advisable.

Even Galileo didn't fully appreciate what he had at first. In his letter to the doge accompanying the gift of the instrument, he wrote:

> This is a thing of inestimable benefit for all transactions and undertakings, maritime or terrestrial, allowing us at sea to discover at a much greater distance than usual the hulls and sails of the enemy, so that for 2 hours and more we can detect him before he detects us and, distinguishing the number and kind of the vessels, judge his force, in order to prepare for chase, combat, or flight; and likewise, allowing us on land to look inside the fortresses, billets and defenses of the enemy from some prominence, although far away, or also in open campaign to see and to distinguish in detail, to our very great advantage, all his movements and preparations; besides many other benefits, clearly manifest to all judicious persons.

Among the benefits *not* immediately manifest, presumably, were those that such an instrument might confer on a professor of mathematics—that is, an

astronomer. Still, if observation of the night sky wasn't
the first use to come to mind, it nevertheless would have
provided an irresistible temptation for someone with
Galileo's curious and obstinate disposition, and over the
next few months he began to act on it. During a visit
to his native Florence that October, for instance, he
demonstrated the instrument to a former pupil of his,
Cosimo de' Medici, the current grand duke of the
Tuscan court, and together they marveled at the earthlike irregularities on the surface of the Moon. During
the late summer and autumn months of 1609, Galileo,
along with an assistant in his shop, continued grinding,
polishing, lengthening, and otherwise pondering and
improving the pieces of the instrument—its leaden tube
and two lenses—and by late November he had completed a model capable of magnifying twenty times. It
was this instrument with which he initiated his first formal campaign of observations, complete with pad, pen,
and wash, and the moment he did, the instrument
became something more than a toy or a maritime aid.
What artists had been advocating was nothing less than
a new way of looking at the world, and now here it was,
only more: a new way of looking at the universe.

Over the preceding two centuries, Europe had awak-
ened, as if from a long winter's nap, and as innocent as
Adam. Explorers had blinked in the fresh light, and then
they had set out, voracious for knowledge and experience, until they had devoured the globe. There was
nowhere to go but up.

On November 30, 1609, Galileo carried his *perspi-*
cillum and artist's utensils into the garden behind his

apartment in Padua and began to study the Moon. The first time he'd observed the night sky a couple of months earlier, he might have been harboring grave doubts about the wisdom of the ancients, especially in astronomical matters, but he would have had no reason to anticipate finding anything through this tube that could influence his opinion or anyone else's, one way or the other. He simply would have been exercising a natural curiosity. Still, for someone of Galileo's professional background and personal disposition, one look would have been enough. He had raised his new instrument toward the night sky and understood at once that there was more to see—and more to seeing—than meets the eye.

CHAPTER TWO

✳

God's Eye

I n 1609 the world—which is to say, that portion of the universe at the center of which inhabitants of Europe, Asia, Africa, and the Americas might imagine themselves—was innocently living out its final year of isolation. According to standard illustrations of the day, the universe looked like this: Draw nine concentric circles. The circle in the center, of course, was the Earth, and surrounding it in outward order were, first, the orbits of the seven wandering celestial bodies—Moon, Mercury, Venus, Sun, Mars, Jupiter, and Saturn—and then, defining the farthest reaches of the universe, the Sphere of the Fixed Stars.

The next year all that changed. In March 1610, barely three months after making his first formal observations of the night sky, Galileo published his findings under the title *Sidereus Nuncius*. He intended the title to suggest a "message from the stars," but from the start translations tended to favor a "messenger from the stars" (in English, either *The Sidereal Messenger* or *The Starry*

Messenger), heralding the arrival of not only the news but its bearer, a heavenly intermediary of sorts—the pamphlet perhaps, or the professor of mathematics from Padua, or the astonishing new *instrumentum* itself.

However ambiguous the Latin on the title page, the language within was unusually straightforward. What Galileo had to report needed no rhetorical help, none of the flourishes or embellishments that marked most scholarly writing of the period. He might indulge in the occasional flight of self-congratulation, letting the reader know that in improving his instrument, he had spared "neither labor nor expense," or that he was the first to witness phenomena "never seen from the creation of the world up to our own time," or that he'd succeeded where the greatest minds had failed, settling "all the disputes which have vexed philosophers through so many ages." But when he came to the revelations themselves, he allowed the accumulation of evidence to speak for itself—as if, as he himself wrote, "we are at last freed from wordy debates."

First he introduced the instrument to his readers, a spyglass "by means of which visible objects, although far removed from the eye of the observer, were distinctly perceived as though nearby." Next he addressed the Moon: the "large or ancient spots" that are visible with the naked eye, then the spots "smaller in size and occurring with such frequency that they besprinkle the entire lunar surface," the "uneven, rough, and very sinuous line" that divides the sunlit regions from the dark, the "very many bright points" that "appear within the dark part of the Moon," and his own estimation, based on measurements made through his observations, that the heights and depths of the Moon's "protuberances and gaps" surpass those on Earth. Then he addressed the newly visible stars—in Pleiades, for instance, where the known population grew from six stars to forty, or in Orion, a constellation suddenly five hundred objects richer, or in the galaxy itself, the ancient milky spill across the night sky that now turned out to be "nothing but a congeries of innumerable stars grouped together in clusters. Upon whatever part of it the *occhiale* is directed, a vast crowd of stars is immediately presented to view." Finally he came to the discovery that was easily the most momentous in his book.

On January 7 he had noticed three stars he'd never seen before. They didn't seem especially noteworthy at the time, though that night he'd happened to mention them in a letter as an example of the kind of phenomenon he was turning up with every round of observation: "and only this evening I have seen Jupiter accompanied by three fixed stars, totally invisible because of their smallness." He himself couldn't say why—perhaps it was because of the striking arrangement of these three stars, all in a row and all in line with Jupiter—but the following evening he returned to them. The first time he saw the three stars, two had appeared to the east of Jupiter, the other to the west; now he found all three to the west. Had Jupiter somehow strayed off course, wandering to the east when according to all astronomical observations and predictions, it should have been moving west? The next evening was cloudy, but on January 10 he resumed his observations, and this time only two of the stars near Jupiter were visible, both to the east, the planet itself presumably obscuring the third. The following night he found the same general arrangement, though the distances from Jupiter were different, as were the apparent sizes of the stars. For several days he had been trying to calculate how Jupiter could be traveling in such a way as to create these myriad arrangements. Now it was beginning to dawn on him that it might not be Jupiter that was moving, but the new stars themselves.

Over the following days he found himself returning to this possibility. On the twelfth he observed that not only were two stars visible, one on either side of Jupiter, but this: "The westerly star was a little smaller than the easterly, and Jupiter was in the middle, distant from either star by about as much as its own diameter; and perhaps there was a third [star], very small and very close to Jupiter—or, rather, there really was, I having observed with great diligence and the night being darker." The next night brought a surprise: a fourth star. "Three were on the west and one on the east," he noted. "All these stars displayed the same size, and although small they were nevertheless very brilliant and much brighter than fixed stars of the same size.

"On the fourteenth, the weather was cloudy."

Gradually, as he'd grown accustomed to the subtleties of what he was seeing, Galileo had begun to refine his observations. At first he had noted only absolutes: the quantity of stars and whether they were east or west of Jupiter. Then he'd started noting their sizes relative to one another and their comparative dis-

tances to Jupiter. On the fifteenth he added a further
refinement: the times of observations. And so, "in the
third hour of the night," he found four stars again, all to
the west, all in a row. "But in the seventh hour," he went
on, "only three stars were present in this arrangement
with Jupiter." Between these two observations he had
taken a significant step. He had opened his journal and
begun copying the previous week's worth of notes; he
also had switched from the informality of Italian to the
scholarly Latin. There was one possible explanation for
what he was observing, he now thought. The four
objects weren't simply stars but planets, wanderers in
their own right, and weren't simply wanderers, but
moons of Jupiter.

Galileo had been entertaining the idea of writing an
account of his observations, but this realization was *it:*
a reason to establish his priority by rushing the manuscript into print as soon as possible, certainly, but also
the single piece of news that transcended everything
else he'd observed through his *perspicillum.* Here was
the one discovery that appeared not only to contradict
the teaching of ancient authority but to reverse it,
almost.

Like the other observations he'd made, this was in-
conclusive. Nearly seventy years earlier the astronomer
Nicholas Copernicus had published a lengthy mathematical argument supporting a model of the universe
with the Sun at the center. Since then one of the main
criticisms that had arisen against this model was that it
required the Moon to orbit the Earth, and the two bodies together to orbit the Sun. Why should the Earth be

the only planet to have a moon? For that matter, why should the universe have *two* centers of rotation, the Earth and the Sun? If, however, Galileo's contention were correct—that Jupiter shared with Earth the distinction of being the host to at least one moon—it might prove just persuasive enough to revive the debate over the Copernican model of the universe, and even to tip the balance.

The modest packet of twenty-four pages created an immediate sensation. Even before publication, and even on the far side of the Alps, a banker in Augsburg who regularly conducted business with the Jesuit College in Rome wrote to the senior mathematician there asking whether the rumor he'd heard was true: that a mathematician in Padua had "discovered four new planets, new to us, having never been seen, as far as we know, by a mortal man, and many fixed stars, not known or seen before, and marvelous things about the Milky Way." The next day, March 13, 1610, *Sidereus Nuncius* was published, and that same day an English visitor to Venice, Sir Henry Wotton, forwarded a copy to his king along with a letter that read in part:

> Now touching the occurrents of the present, I send herewith unto his Majesty the strangest piece of news (as I may justly call it) that he hath ever yet received from any part of the world; which is the annexed book (come abroad this very day) of the Mathematical Professor at Padua, who by the help of an optical instrument (which both enlargeth and

approximateth the object) invented first in
Flanders, and bettered by himself, hath discovered four new planets rolling about the
sphere of Jupiter, besides many other unknown fixed stars.

The five hundred copies of the book sold out immediately; Galileo himself received only ten of the thirty
for which he had contracted. In Florence two weeks
after publication a courier arrived from Venice with a
package, and a crowd quickly gathered, assuming the
contents to be a spyglass. It turned out to be a copy of
Sidereus Nuncius instead, and the onlookers refused to
disperse until they'd heard every word. In Venice itself
a similar package attracted a similar crowd, and this
time they were rewarded with a spyglass, which, to the
beleaguered owner's dismay, they monopolized for
hours.

At once it was apparent that if what he'd written
turned out to be true, Galileo had exceeded any achievement in the Age of Discovery. As the British natural
philosopher William Lower wrote on hearing the news,
"Me thinks my diligent Galileus hath done more in his
three fold discouerie than Magellane in openinge the
streightes to the South sea." The Scottish poet Thomas
Seggett added,

Columbus gave man lands to conquer by blood-
    shed
Galileo new worlds harmful to none. Which is
    better?

Johannes Faber, a prominent medical and botanical philosopher, proclaimed:

> Yield, Vespucci, and let Columbus yield. Each of
> them
> Holds his way through the unknown sea, it is true.
> But you, Galileo, alone gave to the human race the
> sequence of stars,
> New constellations in heaven.

Where the ancients had detected perfection, Galileo found mountains and valleys. Where anyone else located a whitish wash, Galileo saw stars. Where the fates had fixed the number of celestial wanderers forever at seven, Galileo added four more. As Sir Henry Wotton had written in that same day of publication letter, in one slim volume Galileo and his instrument had "overthrown all former astronomy." Not that Galileo was ready to make this claim explicitly—not yet, anyway—and not that anyone would have accepted it outright. As the banker in Augsburg wrote, "I know very well that 'to believe slowly is the sinew of reason,' and I have not made up my mind about anything."

Some refused to look. The planets wouldn't be there because they couldn't be there. One astronomer in Florence argued that "these satellites of Jupiter are invisible to the naked eye, and therefore exercise no influence on the earth, and therefore would be useless, and therefore do not exist."

Some looked and refused to see. Critics accused Galileo of planting his "planets" in the spyglass. One adopted the hypothesis that while yes, the surface of the Moon is mountainous, a transparent covering as high as the highest peaks preserves the purity of the planet's spherical perfection.

Some looked and *couldn't* see. According to one eye-witness at a demonstration Galileo performed in Bologna in April 1610, none of the several learned men present could see the two moons of Jupiter that Galileo claimed to see (and that Galileo recorded in his own log of observations for the evening). "On Earth it works miracles," this observer wrote; "in the heavens it deceives, for other fixed stars appear double." He called the planets "fictitious," reported that "all acknowledged that the instrument deceived," and concluded that "the wretched Galileo" had to slink away in disgrace "because, full of himself, he hawked a fable."

Then there was the simple act of seeing—an act that suddenly was seeming not so simple after all. "In this way," Galileo had written in the opening paragraphs of *Sidereus Nuncius,* describing the great advantage of observing through an optic tube, "one may learn with all the certainty of sense evidence." But how much certainty did "sense evidence" actually offer?

Observers had been using tubes without lenses to study the heavens for hundreds of years, and even Aristotle, in his *Generation of Animals,* had noted the salutary effects of a tube in clarifying a distant image: "The man who shades his eyes with his hand or looks through a tube will not distinguish any more or any less the differences in colors, but he will see further; at any rate, people in pits and wells sometimes see the stars."

He took this observation to its logical conclusion: "[D]istant objects would be seen best of all if there were a sort of continuous tube extending straight from the sight to that which is seen, for then the movement which proceeds from the visible objects would not get dissipated; failing that, the further the tube extends, the greater is bound to be the accuracy with which distant objects are seen."

A spyglass, of course, worked partly on the same principle, but the introduction of glass into the tube didn't just magnify the image and gather more light than could an eye and tube alone. It also produced distortions—elongations, blurriness, color fringes. Galileo, for one, quickly learned how to combat some of the problems. In order to overcome the unsteadiness of his grip and the pulsing of his blood, he secured the instrument to a stable surface. He figured out that by lengthening the tube, and therefore the distance between the lenses, he could see near objects more clearly; conversely, by shortening the tube, he could better see distant objects. By early January 1610 he had also hit upon the notion of stopping down the object glass, the lens at the end of the tube away from the observer, by grinding a lens larger than he needed, then placing cardboard over it so as to leave an oval opening in the center. In that way light would be passing through the area of the lens surface with the least curvature and therefore the fewest distortions. Even so, of the more than one hundred optic tubes that Galileo produced, he considered only ten truly representative of his powers and able to afford a view of the moons of Jupiter, and among those, even the finest examples of Galileo's craftsmanship yielded only a small fraction of the Moon's surface.

Most observers, of course, didn't have access to an optic tube from the studio of the master. The most politically influential did—Galileo made sure of that, through regular mailings to the crowned heads and other potential advocates or patrons—but all others in those first formative years had to satisfy their immense curiosity with inferior instruments that introduced even greater distortions into the observations. According to an account by one German professor, "the body of Jupiter was seen completely on fire, so that it appeared separated into three or four fiery balls, from which thinner hairs were spread in a downward direction, like the tail of a comet."

In astronomy the reliability of an observation had always depended on the quality of the instrument or the observer's vision. The spyglass, however, introduced uncertainties about the evidence of the senses that engaged not just the eye but the mind. Even an observer willing to accept the instrument as fundamentally trustworthy still had to confront the question of what it actually did—not so much how it worked, though the subject of optics certainly held promise as an area of investigation, as what it revealed, that is, how to interpret its momentous revelations, or even *whether* to interpret them.

Without prior experience, observers had no way of knowing for sure what they were seeing. During one demonstration Galileo showed by daylight the inscription above the entrance to a building several miles away, then by night the satellites of Jupiter. In so doing, he made a point that needed making: An instrument that saw truthfully on Earth surely did the same in the heavens. But not even Galileo could anticipate what form any one heavenly truth might take. Only through repeated observations and independent confirmations could an interpretation even begin to approach a degree of certainty. The interpretation of Jupiter as four big fireballs didn't endure only because nobody saw it that way except for one poor German professor, not because it was any less credible than the other celestial phenomena first coming into view.

The spyglass, then, presented a challenge to even the most sympathetic or expert observer simply because it was the first instrument to extend one of the human senses. Unlike spectacles or magnifying lenses, the optic tube offered not just a distortion of what was already there, but *more*. It revealed evidence that was different from what the naked eye could see, evidence that *wasn't otherwise there*. And if you raised it toward the sky, it revealed—with "all the certainty of sense evidence"—evidence that contradicted what *was* otherwise there. And the two views of the universe couldn't both be right.

The evidence of the senses had always sufficed. The reason that a concentric-circle view of the "Kosmos," from the Greek for "order and harmony," had survived for two thousand years was that it made sense, as in common sense, as in what was readily observable through the sense of sight. It placed a stationary Earth at the center of Creation and set the heavens spinning.

Not that there weren't irregularities. The planets, for instance, didn't always exhibit the uniform circular motion that one might expect—indeed, that Greek dogma demanded—from a heavenly body. Sometimes a planet would be progressing night by night from west to east through the zodiac and then suddenly, over a period of nights, would back up, going from east to west, before eventually resuming the usual west to east motion. Partly for this reason Aristotle, among others, had hypothesized that the celestial orbs travel along interlocking transparent spheres, one inside the next, one each for the Sun, Moon, and every planet, a final outer sphere for the fixed stars, and each the domain of an invisible mover. These spheres alone, however, couldn't account for all the movements, so he and other philosophers and mathematicians added other adjustments—spheres within spheres, spheres tangential to spheres—until they'd produced fifty-five in all, each with its own guiding spirit.

These adjustments helped to account for observed phenomena—or, as mathematicians of the day would say, to "save the appearances." But it wasn't until Ptolemy—the same ancient whose *Geography* had proved so inspirational, if ultimately so unsatisfying, to the explorers of the fifteenth century—that this system found a thorough basis in mathematics. In the second century A.D. Ptolemy observed the heavens at length, found himself agreeing with the Aristotelian model of the universe, and constructed for it a complete mathematical accounting of all the celestial motions. He called this work *The Mathematical Collection,* but

Arabic translators of the ninth century, in deference to his unprecedented achievement, called it "The Greatest Composition"—*Al-mageste,* or, as it came to be known, *Almagest.* As a result of Ptolemy's herculean efforts, the Aristotelian system of the universe now matched the evidence of the senses *and* worked mathematically. He'd even pared the number of spheres down to forty.

Further mathematical improvement didn't arrive for thirteen hundred years. Nicholas Copernicus was born in 1473, at the height of the Age of Discovery; the news of a New World reached him in early adulthood. Still, Copernicus was cautious by nature. He never intended to challenge the status quo, merely to amend it. He began formulating *De revolutionibus orbium coelestium,* or *On the Revolutions of Celestial Orbs,* more than thirty years before he published it, in 1543, and even then he did so only because of the urgings of friends who admired its superior mathematics and its novel interpretation of the visible universe. What originally prompted him to explore the question of how the heavens work, and what ultimately motivated him to make his answer public, were less a passion for an alternative system that might work than a dissatisfaction with the one that didn't.

The universe had gotten ungainly. Mathematicians had created spheres and circles called epicycles, equants, eccentrics, deferents. These they had added and subtracted and interchanged and appended, one to another, with an abandon that verged on caprice. Yet, Copernicus argued, mathematicians had failed fundamentally. They *hadn't* saved the appearances. By the sixteenth century-

the calendar and the seasons diverged by weeks; predictions of celestial phenomena such as eclipses and con-
junctions could miss their marks by a month.

In the end, though, Copernicus's foremost objection
to the Ptolemaic system and its numerous variations
was aesthetic. Mathematicians, he complained in his
preface to *On the Revolutions,* had been unable "to discern or deduce the principal thing—namely the shape of
the Universe and the unchangeable symmetry of its
parts. With them," he concluded, "it is as though an
artist were to gather the hands, feet, head and other
members for his images from diverse models, each part
excellently drawn, but not related to a single body, and
since they in no way match each other, the result would
be monster rather than man."

The mathematical component of his work was widely
praised and adopted. For the first time a competing cosmology was at least as mathematically valid as
Aristotle's, and maybe more so. With the Copernican
system, mathematicians could make predictions about
celestial events with far greater accuracy, and the calendar, after the reform of 1582 instituting many
Copernican innovations, again matched the seasons.

The central conceit of a moving Earth, however, was
another matter. In fact, next to the usefulness of his
mathematics, it was considered to be of little or no con-
sequence, a means that had gotten Copernicus to a far
more productive end. Copernicus was far from original
in suggesting a Sun-centered, or heliocentric, model of
the universe. During Aristotle's lifetime, Heraclides of
Pontus had hypothesized it wasn't the heavens that

made a complete turn every day, it was the Earth, and one hundred years later Aristarchus of Samos even went so far as to suggest it was the Earth that moved around the Sun, not vice versa. But Copernicus was the first to construct a mathematical system as comprehensive as the one Ptolemy had built for Aristotle's Earth-centered, or geocentric, model. He had answered one traditional objection to a heliocentric system, the lack of a mathematical basis. But like Heraclides and Aristarchus, he hadn't provided a better match for the evidence of the senses, and the Aristotelian system would endure as long as it *did*.

"I began to doubt the faith of my own eyes," was how the man who would become Europe's leading astronomer recalled the evening in 1572 that he witnessed the seemingly impossible: a *nova,* or new star. Tycho Brahe, a Danish observer, knew every object in the night sky by heart, yet here, suddenly, was not only a new star in the constellation of Cassiopeia, but one that easily outshone nearly everything else in the heavens. The following year, after it had faded away, he published a brief argument, based on his observations, that the nova belonged not to the presumably changeable region this side of the Moon but to the far more distant, heretofore inviolable sphere of the fixed stars. Five years later a comet blazed across the sky, and once again Tycho determined that a new object in the night belonged to the formerly inalterable realm of the celestial.

By 1588 he was ready to resolve one old debate once and for all. "There are no solid spheres in the heavens," he declared in his *De mundi aetherei recentioribus phaenomenis,* or *Second Book About Recent Celestial Appearances.* The spheres within which the planets traveled in their nocturnal wanderings were symbolic; they and their guiding spirits were merely representative of geometric motions and didn't cause the motions themselves. Although Tycho circulated some copies among friends in 1588, the book wasn't widely published until 1603—as it happened, just in time for the next sphere-shattering event, the New Star of 1604.

There had been reports of novae throughout history. There had been accounts of comets too. But scholars were increasingly reluctant to accept the astrological or theological explanations of portents and miracles that had sufficed in the past, especially when the new instances of celestial phenomena were accompanied by fresh and irrefutable observations and calculations— and *most* especially once the prevailing view of the universe had suffered the loss of its spheres. It was one thing for the Aristotelian Kosmos to yield to Copernicus's system on mathematical grounds; modern correctives to ancient teachings were overturning every kind of scholarly discipline all the time. It was quite another concession for the Aristotelian Kosmos to yield on the appearances themselves.

Tycho himself proposed an alternative to both the Aristotelian and Copernican systems, one that required the planets to revolve around the Sun while the planets and Sun together revolved around the Earth—not an especially felicitous option. Among mathematicians, the growing consensus was that while the Aristotelian system was showing its age, the Copernican hardly offered a useful alternative. What astronomy needed, this argument went, was a bold new hypothesis at least as mathematically sound as the Copernican, if not more so (the predictions of celestial phenomena were still sometimes off by days), but as satisfying to the senses as the Aristotelian—what one leading mathematician called "the radical renovation of astronomy."

Galileo, for his part, was *not* claiming that that was what he'd seen through his spyglass. Galileo was a mathematician, and according to the division of astronomical labors that had held since ancient Greece, mathematicians concerned themselves only with understanding the way the universe works—the evidence of the senses. On the basis of that evidence, they then performed the mathematics sufficient to predict eclipses and other planetary phenomena, or to add an ecliptic and save the appearances.

The *meaning* of those appearances, however, was the province of philosophers. Philosophers concerned themselves with an understanding not of the way the universe works but of man's place in that universe. Clearly, one set of rules applied to what happened on Earth—the terrestrial realm—and another to the heavens—the celestial realm. On Earth, objects moved in straight lines, falling toward the ground as if drawn to the center of the universe—the repository, apparently, of all that was earthlike, which was to say, as one ancient philosopher put it, "the filth and mire of the world, the worst, lowest, most lifeless part of the universe, the bottom story of the house." The celestial realm, by contrast, suggested perfection. Heavenly bodies exhibited perfect form, the circle, and Aristotle had no doubt that if close examination was possible, they'd exhibit perfect content too. Down here dwelled the four terrestrial elements: fire, water, air, and earth, in various combinations. Up there, who knew? Who possibly ever could? Still, Aristotle thought a fifth, solely celestial, and undoubtedly incorruptible and eternal element deserved a name, so he called it quintessence.

In the spring of 1611, as Galileo was preparing to visit Rome for the first time since the publication of his findings, the head of the Collegio Romano, Robert Cardinal Bellarmine, asked his Jesuit mathematicians about Galileo's celestial observations. The mathematicians replied that their own observations with an optic tube agreed with Galileo's. This highly influential support, as well as a reception in Galileo's honor at the *collegio,* provided significant validation for the new instrument and its evidence—that is, for the mathematical concerns. What it pointedly did not provide, in the opinion of Cardinal Bellarmine, was validation of or even preference for a particular model of the universe— the concerns of philosophers. As the cardinal later wrote about the Copernican system, "To demonstrate that the appearances are saved by assuming the Sun at the center and the Earth in the heavens is not the same thing as demonstrating that in truth the Sun *is* in the center and the Earth *is* in the heavens. I believe that the first demonstration may exist, but I have very grave doubts about the second; and in case of doubt one may not abandon the Holy Scriptures as expounded by the holy Fathers."

At first, Galileo adhered to the narrow province of the

mathematician. For his discovery of the satellites of Jupiter in *Sidereus Nuncius,* for instance, he presented painstaking documentation: every illustration from every observation from every night, and sometimes several times a night, between January 7 and March 2, the latest date he could make observations and still include them, and a mere eleven days before publication. Partly because of the constraints of time, partly so as not to diminish the impact of his findings, but primarily because they fell outside his discipline, he declined to give philosophical interpretations.

Still, the strain showed. "Let these few remarks suffice us here concerning this matter," he writes at one point about the Earth. "For we will demonstrate that she is movable and surpasses the Moon in brightness, and that she is not the dump heap of the filth and dregs of the universe, and we will confirm this with innumerable arguments from nature." In the end, however, the only philosophical interpretation to which he was willing to commit himself was an offhand remark, "as all planets go around the sun," a contention technically consistent with either the Tychonic or Copernican model of the universe.

During the same visit to Rome at which the Collegio Romano officially verified his observations, Galileo received another accolade from a group with a far different purpose, the Accademia dei Lincei, or the Academy of the Lynx-Eyed. Unlike the Jesuits, who had validated only the mathematical significance of spyglass observations, the Linceans warmly endorsed not only the philosophical implications but the mathe-

matician's right to explore them. It was at their dinner
in honor of Galileo, on April 14, 1611, that a Greek poet
and theologian, John Demisiani, bestowed on the instrument the Italian name *telescopio,* from the Greek for "to
see at a distance," and it was a measure of Galileo's
feeling of fellowship with the group that this is the term
he thereafter adopted almost exclusively.

The division between mathematicians and philosophers was, to Galileo, like-minded Linceans, and other
adventurous intellects of the day, a false one. The deference that philosophers accorded ancient wisdom
seemed slavish, an unquestioning allegiance to what his
great mathematician contemporary Johannes Kepler
once called, and Galileo echoed in his own writing, "a
world on paper." Aristotle himself, Galileo liked to
argue, wouldn't have held the positions his followers
ascribed to him if he'd known of the new evidence, particularly that available through the telescope. The time
had come for mathematicians to philosophize—to save
the appearances *and* to explain them. In a pointedly
symbolic gesture, when Galileo received his appoint-
ment to the Tuscan court in 1610 (he'd paved the way
by naming the moons of Jupiter after the Medicis), he
wrote to the grand duke with a special request: "Finally,
as to the title and scope of my duties, I wish in addition
to the name of Mathematician that his Highness adjoin
that of Philosopher." His wish was granted.

Even so, it wasn't until 1613, in his *Letters on
Sunspots,* published through the Accademia dei Lincei,
that Galileo finally made a public declaration in favor of
one model of the universe over another. By then not

only had the existence of the moons of Jupiter received widespread independent confirmation, but Galileo had further observed that Venus goes through phases, just like the Moon—a physical impossibility under the ancient Earth-centered Kosmos. The Aristotelian spheres already had been shot full of holes by various comets and novae, but the phases of Venus provided a final repudiation of the Ptolemaic mathematical under-pinnings. "An understanding of what Copernicus wrote in his *Revolutions,*" Galileo concluded, "suffices for the most expert astronomers to ascertain that Venus revolves about the sun, as well as to verify the rest of his system."

Mathematicians had known for seventy years that Copernicus saved the appearances of a Sun-centered universe more thoroughly than Ptolemy had saved the appearances of an Earth-centered universe. This was useful information for mathematicians, but useless for philosophers, at least as long as the Earth-centered uni-verse continued to match the evidence of the senses. Even when the crystalline Aristotelian spheres vanished for all time, the question remained: Why abandon a sys-tem in which heavenly objects did what they appeared to do—circle the Earth—for one in which they did not? Why adopt a system in which, for instance, of the two objects that cross the sky every day, Moon and Sun, one actually is moving, and the other isn't? Now Galileo was claiming he had the answer: because it is true. Because the Copernican system is the mathematical model that most closely matches the actual universe. Because the Sun-centered system doesn't match the evi-dence of the senses—*and it doesn't matter.*

As radical as anything Galileo and other Copernicans claimed to see through the spyglass was this assertion itself. Like the backdrops that vanished with the introduction of perspective into art two centuries earlier, the barrier between the celestial and the terrestrial realms didn't fully reveal what it was doing there until it wasn't. Now it was gone, and in its absence came a recognition of what role the evidence of the senses had played in the understanding of the universe.

Galileo himself would recognize the limitations of sense evidence. By July 1609, not long after Galileo would have first heard about the marvelous new invention from the north, the English scientist Thomas Harriot already had secured for himself a "perspective cylinder" and had begun observing the Moon. He noticed shadings; he sketched a drawing. Yet it wasn't until the following July, after he had acquired a copy of *Sidereus Nuncius,* that he understood what it was he'd seen a year earlier: mountains and valleys. William Lower, a friend and fellow scientist, wrote to Harriot of his own similar edification in the wake of Galileo's renderings: "[I]n the moone I had formerlie observed a strange spottednesse al over, but had no conceite that anie parte thereof mighte be shadowes." Actually, what Lower was now calling "a strange spottednesse" he previously had characterized somewhat more colorfully. "In the full," he'd written of the Moon a year earlier, "she appears like a tart that my cooke made me last weeke; here a vaine of bright stuffe, and there of darke, and so confusedlie all over. I must confess I can see none of this without my cylinder."

Galileo, of course, had no prior experience with

observing through a spyglass, either, yet he managed to sketch celestial bodies so convincingly that not only did his renderings immediately become the defining ones, but so did his method. In *Sidereus Nuncius* he'd written about the surface of the Moon that "its brighter part might very fitly represent the surface of the land and its darker region that of the water. I have never doubted that if our globe were seen from afar when flooded with sunlight, the land regions would appear brighter and the watery regions darker." After this early observation and speculation, however, he began exploring other options—and dropped the idea. "I do not believe that the body of the Moon is composed of earth and water," he wrote in 1616, and several years later he explained why: "If there were in nature only one way for two surfaces to be illuminated by the sun so that one appears lighter than the other, and that this were by having one made of land and the other of water, it would be necessary to say that the moon's surface was partly terrene and partly aqueous. But because there are more ways known to us that could produce the same effect, and perhaps others that we do not know of, I shall not make bold to affirm one rather than another to exist on the moon."

The difference in quality of instruments no doubt helped, but so did the difference in the quality of the observer. Galileo insisted on differentiating between speculation and assertion, between hypothesis and theory. His idea of the investigative process didn't end with sense evidence, nor did it end with philosophizing. Instead it began with both, infused them with reason, and ended only after examining the alternatives.

By his own standards, the ones he'd employed in *not* drawing a conclusion about the presence of seas on the Moon, Galileo's reasoning regarding the phases of Venus as proof of the Copernican model wasn't strictly logical. The Tychonic system also predicted the phases of Venus; even the often worshipful Kepler criticized Galileo for this seeming blunder. But Galileo had come to understand that final and complete proofs weren't always possible, given the available evidence; sometimes the best one could do was construct a model that accounted for all the available evidence, test the model rigorously, and, with each test it passed, move closer to accepting it. Seas on the Moon couldn't be tested, so he abandoned that hypothesis; Copernicanism could, and after years of testing, Galileo accepted it.

In so doing, he hadn't quite confirmed through observation the hypotheses of Copernicus, and he had done nothing to satisfy the math of the Copernican system, but he *had* changed the terms of the debate. While everyone around him debated about the right of a mere mathematician to pronounce on philosophical matters, about the right of a scholar to assert the priority of an unproved personal interpretation over received theological truth—about whether or not to part the proverbial curtain—they failed to notice that the terms of the debate, the assumption they all were willing to accept at the outset of the argument, had changed: The curtain was there.

He had to look. For Galileo, what lay beyond the opaque curtain of blind faith—of a universe divided between the celestial and the terrestrial, between Heaven and Earth, between there and here, that and this—was the constitution of the cosmos itself. The telescope allowed man's senses to roam where he himself could not. The answer to the question of which mathematical model matched the workings of the universe wasn't forever out of reach. It was there for the asking—for the taking. Who, if promised a glimpse of God, wouldn't part a curtain?

"O telescope," wrote Kepler, "instrument of much knowledge, more precious than any sceptre! Is not he who holds thee in his hand made king and lord of the works of God?" The Scottish poet Thomas Seggett agreed, glorifying Galileo for making gods of mortals by revealing hidden planets and pointing out that while Galileo might owe much to God, Jupiter owed much to Galileo. In France, Queen Marie de Medici, unwilling to wait until her servants had finished attaching a spyglass to a window, fell to her knees in the presence of her subjects, astonishing and frightening them. And during his April 1610 audience in Rome, Galileo himself obediently knelt before the pope, who dispensed with precedent and commanded him to rise.

Through the telescope man had completed the intellectual journey he had begun two centuries earlier with the introduction of perspective into art. The telescope bridged Earth and sky in a manner both literal and metaphorical. What had been terrestrial was now celestial; what was celestial, terrestrial. Earth moved through the heavens, one more wanderer, and the once perfect heavens suffered the same fate as Earth. The perspective cylinder united mathematics and philosophy, astronomy and physics, sense evidence and geometry, the ancient world and the modern, Creation and Creator. In two centuries man had gone from being the apple of God's eye to being God's eye.

"They go about defending the inalterability of the sky, a view which perhaps Aristotle himself would abandon in our age," Galileo wrote of philosophers in *Letters on Sunspots,* anticipating criticism for his revelation that even the Sun suffers imperfections.

> Well, if alteration were annihilation, the Peripatetics would have some reason for concern; but since it is nothing but mutation, there is no reason for such bitter hostility to it. It seems to me unreasonable to call "corruption" in an egg that which produces a chicken. Besides, if "corruption" and "generation" are discovered in the moon, why deny them to the sky? If the earth's small mutations do not threaten its existence (if, indeed, they are ornaments rather than imperfections in it), why deprive the other planets of them? Why fear so much for the dissolution of the sky as a result of alterations no more inimical than these?

Galileo knew why. "These men are forced into their strange fancies by attempting to measure the whole universe by means of their tiny scale," he continued. "Our special hatred of death need not render fragility odious. Why should we want to become less mutable?"

Galileo grew old, went blind, died fearing that his

knowledge would perish with him. "My life wastes away," he once said, "and my work is condemned to rot." Not so, however, the evidence on which he'd based his work. "The spyglass is very truthful," Galileo wrote early on, "and the Medicean planets are planets [that is, wanderers, not fixed stars] and, like the other planets, will always be." Whatever he saw through the telescope was here to stay. His interpretations might not endure, his hypotheses might not endure, his standards and tests and conclusions might not endure, but the evidence itself would. It would outlive him because it resided not in him but in his instrument. The center of mathematical and philosophical investigations—and with it, the very center of the universe—had shifted, once and forever, from the eye of the beholder.

Part II

*

BELIEVING

✳

CHAPTER THREE

By the Numbers

It was a bad day for looking into the Sun. November 7, 1631, had started cloudy—or, rather, had *continued* cloudy. For two days rain had been falling in Paris, and Pierre Gassendi, keeping vigil in a dark room, had no reason to think that at any time in the near future the clouds would part. But shortly before nine in the morning, they did, briefly. The rain stopped, the clouds withdrew, the Sun came out—just long enough for him to get a glimpse at the image of it that he was projecting on a white sheet across the room. He bent close to the image of the disk, eight inches in diameter, and studied it for signs of the planet Mercury. All he found, however, was a tiny dot—a sunspot, perhaps. At any rate, it was too small to be Mercury. Dutiful observer that he was, Gassendi noted the position and dimensions of the dot on the disk—one never knew what might prove useful later—and resumed his vigil in the dark.

The coming transit of Mercury—the passage of the planet, from the perspective of an observer on Earth, across the surface of the Sun—had attracted the attention of astronomers throughout Europe. The previous year, and the year before that, the great astronomer Johannes Kepler had urged fellow astronomers to observe this relatively rare celestial event, and Gassendi had prepared for it as he would for any other observation of the Sun. He had cut a single hole in a window shade, positioned a telescope so that it would funnel the light that passed through the opening, and arranged a white sheet to catch the projected image. On that sheet he'd traced a circle that would precisely fit the image of the Sun's disk. On this circle he had drawn two lines through the center—two diameters, at right angles to each other. Then he'd marked the lines, dividing each diameter into sixty equal segments. According to his calculations, the disk of Mercury, should it appear, would occupy about four of those segments and complete its passage across the circle in a matter of hours.

But the weather wasn't cooperating. The Sun disappeared behind the clouds. It reappeared, then disappeared again. When it appeared for the third time that morning, Gassendi once more examined the circle on the white sheet, but the only change he noticed was the same one he'd been following all day, the position of the tiny dot. It was still too small to be Mercury, but it was also moving too fast to be a sunspot. Gassendi knew from previous observations of sunspots that the Sun completed one rotation on its axis every twenty-six days, so the distance a sunspot would travel across his circle in one day should be around four diameter segments. But this "sunspot" was covering that territory, and more, in a matter of hours.

By the time he realized that what he was seeing must be Mercury, it was almost too late. He managed to get only one more reading. Still, at the end of the transit, Gassendi was left contemplating a result that showed, as one colleague put it, an "entirely paradoxical smallness." When he compared the disk of Mercury with the diameter marks on the paper, Gassendi found it "could scarcely exceed two-thirds of one division"—a mere sixth of the diameter that he had expected.

The lessons of the 1631 transit of Mercury (for those few who even recognized it as an opportunity for insights into the future of the telescope) were twofold. First, it clearly demonstrated the need to use telescopes in astronomical observations. Kepler's published admonitions to observe the transit had suggested that Mercury would be equally evident with or without the help of a telescope, but those observers who took him at his word and tried to use only their eyes wound up seeing nothing.

Second, the transit demonstrated a need for greater accuracy in astronomical measurements. As it turned out, even those observers who did use telescopes couldn't have known what to expect or even when to expect it. Kepler had predicted the transit for the seventh of November, but suggested starting on the sixth and continuing to the eighth, just in case. Gassendi took no chances—he knew the state of the art of astronomical predictions—and began his vigil on the fifth. In the end the transit *had* occurred on the seventh; at least astronomy could get that much right. But what he found was so different from what astronomers expected that when Gassendi published his results a year later, one colleague warned that critics "will doubt from this whether you have really seen Mercury himself."

The colleague was right. Gassendi's astonishment, bordering on incomprehension, was shared by astronomers across Europe. "It is indeed like that," wrote one Gassendi sympathizer, "how often we are held by prejudice, so that we either do not admit what is before our eyes or bow as much as possible to a preconceived opinion because of the perverse wont of human nature. Nor do I except myself from that weakness; I would have thought the same, had I succeeded in observing Mercury." It wasn't until eight years later that another astronomer, observing a transit of Venus and determining that the apparent diameter of that planet was less than a third of what astronomers expected, settled the question of what Gassendi had witnessed in November 1631: "Congratulate us, Gassendi, on clearing from suspicion your observation of Mercury, and let astronomers cease to wonder at the surprising smallness of the least of the planets, now they find that the one which seemed the largest and brightest scarcely exceeds it. Mercury may well bear his loss since Venus sustains a greater."

The difference was one of dimensions. The planets that astronomers were expecting were planets that belonged to Ptolemy's mathematical Kosmos; the scale of the universe they were using was a scale that hadn't changed in well over a thousand years. Now, clearly, those dimensions no longer would do, and astronomers were beginning to recognize the need for new dimensions that would—specifically, the scale of the new,

Copernican universe: the sizes of its planets, their distances to the Sun, and how far were the stars.

The desire for a finer degree of accuracy was also what had driven the great astronomers of the previous century, Copernicus and Tycho, both of whom had acted out of frustration with Ptolemy's incorrect data. And they had succeeded brilliantly, one capturing the cosmos in mathematical terms, the other in observational terms, and both with far more refinement than any other astronomers in history. But already their results were proving to be inadequate and in need of reform. For a new generation of astronomers who adopted as their mission the correction and completion of the observations of the past, the difference between themselves and every one of their predecessors dating back to ancient Greece was simple: Now they could call upon the telescope.

Or could they? In the first few decades after Galileo heralded the astronomical applications of his *perspicillum,* the telescope had hardly changed, and the range of its observations had barely expanded. In fact, astronomers for the most part *didn't* observe the heavens with telescopes. A telescope might be useful for satisfying one's curiosity and seeing what Galileo saw, or for observing rare celestial occurrences, such as solar and lunar eclipses or the transits of Mercury and Venus, but during its first decades of widespread use the telescope remained primarily a terrestrial tool, performing precisely those kinds of maritime and military tasks that Galileo and the other early champions of the miraculous new instrument had outlined so persuasively to various states generals and senates.

Part of the problem was that the so-called Dutch, or Galilean, configuration of lenses carried a built-in mechanical limitation, which Galileo and the other, earliest observers had quickly exhausted. In seeming to bring a distant object closer and closer, the telescope wound up concentrating on a smaller and smaller field of view—the length and breadth of what the eye could see through the telescope. At a certain point the field of view grew so exaggeratedly narrow that further magnification could provide only diminishing returns. That point was a magnification of approximately twenty times—the point at which the view could encompass barely a quarter of the Moon's visible surface, or maybe all four moons of Jupiter, and a point that Galileo had reached even before he'd written *Sidereus Nuncius*.

The lenses themselves presented further complications: flaws in the glass, blurriness (especially around the edges of a magnified image), color fringes. More accurate methods of grinding lenses and better grades of glass could eliminate some bubbles and scratches, though rarely all. Stopping down the objective lens, as Galileo had done in observing the moons of Jupiter, decreased the blurriness somewhat. As for the color problem, however, nothing seemed to help.

On the ground these imperfections didn't matter so much; the observer presumably would be able to investigate appearances firsthand to figure out what was really out there and what was in the glass. When observing the heavens, however, an astronomer didn't have that option. Saturn, for example: First Galileo had observed the farthest planet from the Sun and thought it to be oddly "three-bodied"—three stars in a straight line, the largest in the middle, and all almost touching— but when he'd returned to it after an absence of two years he found it "solitary." "Has Saturn devoured his children?" he wrote, bewildered, angry. "Or was it indeed an illusion and a fraud with which the lenses of my telescope deceived me for so long?"

The limitations, however, weren't only mechanical. For a long time astronomers were barely bothering to look for fresh phenomena. As Galileo had written regarding a rival's perceived jealousy, shortly after his initial investigations with the optic tube, "it was granted to me alone to discover all the new phenomena in the sky and nothing to anybody else. This is the truth which neither malice nor envy can suppress." And the word of Galileo was gospel. In 1658—half a century after Hans Lipperhey applied for a patent on a "certain instrument for seeing far," and fifteen years after Galileo's death— Christopher Wren wrote that when Galileo first directed a telescope at the sky, "all celestial mysteries were at once disclosed to him. His successors are envious because they believe that there can scarcely be any new worlds left." Even when new worlds from time to time did appear—moons of Saturn, or markings on Jupiter and Mars that rendered them as good as "new"— astronomers didn't adjust their fundamental expectations. Again, Galileo had helped set the terms of the discussion. When word reached him near the end of his life that another astronomer was observing more details of the Moon than he had, Galileo's response was swift and dismissive: "As to seeing infinite inequalities, it is

true, but they are the same that are seen with our telescopes, except a bit more conspicuous thanks to the
magnification."

The distinction was one of degree, not kind. After
Sidereus Nuncius, all other new celestial phenomena
paled. Even without the ghost of Galileo whispering
discouragement, astronomers were bound to find it difficult to conceive of information as *really* new unless it
took the form it had in 1610: not just astonishing revelations, but sudden, shocking, sphere-shattering, near-
divine revelations.

Certainly the telescope had provided a stunning affirmation of the methods of the New Philosophy, as its
practitioners had taken to calling it. As the first instrument ever to extend one of the human senses, the telescope had dramatically demonstrated the possibilities of
placing one's faith "not in ancient tomes, but in close
observations and personal consecration," as Galileo had
written. "Man will never become philosopher by worry-
ing about the writings of other men." Rather, the
"proper object of philosophy," he believed, was "the
great book of nature"—the evidence of the senses. It
was because of writings such as these, of course, that
Galileo had to kneel before the Inquisition and recant
the conclusions he'd drawn from the evidence, but it
was too late. Like Galileo, who considered himself a
mathematician who philosophized, the New Philosophers insisted on erasing the distinction between practice and theory, between what appearances were and
what they meant. The New Philosophers drew their
conclusions from nature, and so they called themselves

natural philosophers (as distinct from moral philosophers), and they adopted as their creed the motto that
graced the crest of the Royal Society, one of the first of
a rapidly growing number of organizations devoting
themselves exclusively to the exchange of ideas among
natural philosophers: *Nullius in verba*—"On the words
of no man."

From astronomy the lessons of the New Philosophy
had spread to any number of other emerging areas of
study that might reward the intense scrutiny of seemingly familiar objects: botany, geography, geology, min-
eralogy, zoology, physiology, pharmacology. "I profess
both to learn and to teach anatomy, not from books but
from dissections," wrote William Harvey in 1628, "not
from the positions of philosophers but from the fabric of
nature." And just as the telescope had revealed a mysterious new order throughout the heavens, the microscope
was revealing worlds equally elaborate and extensive
within. As Francis Bacon wrote in the Preface to his
1620 New Philosophy call to arms, *Novum Organon,*
"Our only remaining hope and salvation is to begin the
whole labor of the mind again, not leaving it to itself but
directing it perpetually from the very first, and attaining
our end as it were by mechanical aid."

Yet in astronomy itself, after the first round of revelations, the telescope had acquired little additional value
beyond the symbolic. By showing discrepancies
between the old observations and the new—the universe
that the naked eye sees and the one visible through the
telescope—it had created a desire for further information, for more revelations. Still, it had done nothing to

suggest that it might be the way to satisfy that desire. The Galilean telescope had quickly exhausted its inherent limitations, and that, presumably, was that.

Back in 1611, in his *Dioptrice,* the first study of optics inspired by the telescope, Johannes Kepler had suggested a variation that he hoped might improve on the telescope's narrow field of view. In the original Dutch version of the spyglass, he explained, the rays of light that pass through the lens at the far end of the telescope don't come to a focus until they've passed through the eyepiece and come out *behind* it, outside the telescope, near the eye of the observer. Kepler suggested moving the eyepiece back, so that the light rays come to a focus *in front of* it, inside the telescope, and changing the eyepiece from a concave lens to a convex lens, in effect making the eyepiece into a magnifying glass. Let the convex lens at the far end of the telescope magnify the image just as it already did, he suggested; then let a convex eyepiece magnify it *more.*

It worked. The Keplerian telescope offered a much broader field of view. But it also inverted the image. Because the light rays passed through only one lens, what an observer saw in looking through a Keplerian telescope was an upside-down view. In 1645 a book by a Capuchin monk reported a variation on Kepler's variation: a telescope with additional lenses inside the tube to reinvert the image. In observing the heavens, this kind of instrument was virtually useless; the extra lenses absorbed too much light and compounded the already troublesome blurriness. On Earth, however, it offered a significant improvement. A Galilean telescope at a distance of five hundred yards might yield a view all of two yards wide; now, as the price list of one leading optician reported, "an entire army of about 7 or 8 thousand may be beheld clearly at one time and be examined very distinctly indeed." The Keplerian telescope with interior lenses was ideally suited to become—and from that day forward *was*—the terrestrial telescope.

The astronomical telescope followed. In its original upside-down–image incarnation the Keplerian telescope had found few followers among astronomers. Except in the study of sunspots—where the telescope would project the image of the Sun on a sheet, thereby inverting it again and rendering it erect—most astronomers ignored it. The terrestrial telescope with its large field of view, however, presented a tantalizing proposition, and it wasn't long before astronomers realized that simply by removing the interior lenses, they could eliminate the dimness and blurriness while preserving the large field of view.

The telescope was right back where Kepler had left it in his *Dioptrice* more than thirty years earlier—convex objective, convex eyepiece, inverted image, and all. What had changed wasn't the instrument but the natural resistance that astronomers had brought to looking at inverted images. Not until a telescope maker in Naples, Francesco Fontana, reported using Keplerian telescopes to see belts on Jupiter and markings on Mars in the late 1630s and early 1640s had astronomers begun to realize that when one looked at a field of stars or the surface of a planet, it didn't make much difference which end was up.

The very idea of developing the astronomical telescope—the very idea of there even *being* a telescope for
exclusively astronomical purposes—was new. But
unlike its Galilean counterpart, the Keplerian telescope
left ample room for improvement in magnification
before it would reach the limits of its much wider field
of view. Opticians and astronomers now applied themselves to understanding how lenses work because even
before they figured out what to do with the astronomical telescope, they had to figure out what it *could* do.

The key to greater magnification, as they knew, was
the ratio of focal lengths between the two lenses in the
tube, and the key to *that* was the curvature of the objective in particular, the lens at the far end that gathered
light into the tube. What the curvature in a lens does is
redirect, or refract, the light as it passes from one side
of the lens to the other. If the curvature is great, so will
be the refraction. The rays will angle severely inward,
and the more severely they bend, the nearer to the lens
they'll converge, or focus. But if the curvature in the
lens is minor, so too will be the refraction. The rays will
angle more gently, and the more gently they bend, the
farther away they'll focus.

Actually they *won't* focus. The rays won't converge
as long as the lens is spherical, and at the time spherical
was the only shape that most opticians could grind, at
least with any skill. As a result, opticians knew that they
weren't going to get rid of the blurriness (or spherical
aberration) and color fringes (or chromatic aberration)
that plagued the astronomical telescope. The best they
could hope for was to reduce them, which meant grind-

ing lenses with less and less curvature. The less curvature, the greater the focal length; the greater the focal
length, the longer the telescope.

Six to 8 feet—that was the length of a good astronomical telescope in 1645. Five years later it was 10 to
15 feet. Ten years after that, 25 feet. Ten years after that,
40 to 50 feet. By 1673 Johannes Hevelius had constructed a telescope 150 feet long on the shores of the
Baltic Sea, too long for metal tubes to bear the weight,
so he resorted to a kind of wooden trough. Then
Christiaan Huygens dispensed with tubes altogether.
The result was the aerial telescope—a tubeless wonder
that consisted of, down at ground level, a convex eyepiece mounted on a wooden support and, up there, in
the distance, at the far end of a taut thread, high atop a
mast, at the uppermost reaches of a system of pulleys
and weights and guy ropes and the men who manipu-
lated them, a just barely convex lens. And why stop
there? One astronomer cheered the coming day when
aerial telescopes would have a focus of 1,000 feet and
human spectators could marvel at the antics of the animals on the Moon.

The long telescopes, however, proved to be as
impractical as they were, well, *long.* In Hevelius's
model, the wood warped, the ropes changed length with
every shift in humidity, and the whole contraption
swayed in the slightest breeze. In the aerial telescope,
any stray light easily washed out the view. In both, the
challenge of lining up an eyepiece and an objective
more than 100 feet apart could last the better part of a
night, wasting precious observation time. Beyond a cer-

tain length, the long telescope—with or without a tube—was proving to be more of a crowd-drawing curiosity than an instrument of celestial investigation. With occasional, notable exceptions, the new knowledge of the heavens was coming from telescopes of rather modest length, long but not *that* long—usually no more than 30 to 40 feet.

In 1655 Huygens used a 12-foot, 50-power telescope to discover a moon of Saturn, which he named Titan. The following year he erected a 23-foot, 100-power model and solved the riddle of Saturn's puzzling appearance: "It is surrounded by a thin, flat ring, nowhere touching." Giovanni Domenico Cassini, an Italian astronomer, found shadows of Jupiter's moons on the sunlit surface of the planet; he followed markings on Jupiter, Mars, and Venus and deduced from their movements that these planets rotate just as the Earth does, and on the same axis. In 1669 Cassini accepted an invitation to join the new Paris Observatory, where he arranged to transform an old water tower into a support for his telescopes, complete with steps to the top and a balustrade so his assistants wouldn't fall off. Over the following two decades he discovered four more satellites of Saturn using telescopes with lengths of 17, 34, 100, and (an aerial model) 136 feet, and he further determined that the ring around that planet appeared to be, in fact, *two* rings, one inner, one outer, and separated from each other by a gap.

Once again the telescope might have appeared to have run out of room for improvement—and in terms of new phenomena, it pretty much had. In the first few decades of its use as an astronomical telescope, the Keplerian model had helped uncover a host of new celestial phenomena, any one of which alone, if this were 1610 and Galileo had discovered it, would have guaranteed *Sidereus Nuncius* a place in history. But now the Keplerian telescope, like the Galilean before it, had reached its technological limit, at least in terms of seeing what was out there.

As it happened, however, the Keplerian telescope had another distinct mechanical advantage over the Galilean model, though one far less immediately apparent than a larger field of view, or even the more subtle reductions in chromatic and spherical aberration. In 1659, in the same *Systema Saturnium* in which he hypothesized that Saturn was surrounded by a ring, Christiaan Huygens pointed out a peculiar property of the astronomical telescope: Because the rays of light converge (or, actually, nearly converge) within the instrument, rather than behind it, an object inserted into the field of view will appear just as distinct as the image itself. Place a measuring device into the telescope, and you can determine the relative distance between two points in the image— the opposite ends of a planet's diameter, or a planet and a star, or two stars.

In fact, Huygens was not the first to notice this property. Twenty years earlier a British astronomer named William Gascoigne had found a spider thread inside his Keplerian telescope and noticed that the strand looked just as clearly defined, just as sharp, as the heavenly bodies, a secret that nearly perished with him, on July 2, 1644, at the Battle of Marston Moor (and one that became public only when a friend recalled it after reading about Huygens's rediscovery of the principle). Gascoigne had used hairs or thin metal strips, Huygens used a tapered strip of copper, and in 1666 the French astronomer Adrien Auzout used two parallel wires, one stationary, one that moved by means of a finely calibrated screw. But in all three micrometers (from Greek words meaning "small measurement"), the result was the same: For the first time telescopic observers could perform precise quantitative tasks.

From two wires that run parallel to each other it was a short conceptual leap to two wires that intersect—the crosshairs of a telescopic sight, developed by Jean Picard in 1667. Whereas the micrometer allowed astronomers to make measurements within the tube of the telescope itself, the telescopic sight worked in conjunction with measuring instruments. Attach a telescopic sight to one of the giant instruments that measure small angles on the sky—a sector, or a quadrant—and you can fix the right ascension and declination of a heavenly object on the celestial vault with as much precision as a longitude and latitude coordinate on Earth.

Dimensions and distances: The measurements afforded by the micrometer and the telescopic sight, when combined with geometry and trigonometry, could go a long way toward answering the two basic questions of astronomy—the same questions that Pierre Gassendi had brought to his observation of the 1631 transit of Mercury and that observers of the night sky had been confronting since the beginning of time: How big? How far?

Even so, the usefulness of the telescope in these matters still wasn't a foregone conclusion. In the 1670s two
great astronomers each undertook a new survey of the
sky with identical intentions: to improve on the most
accurate measurements of the pretelescopic era. One of
these astronomers would be using instruments fitted
with the new technology of telescopic sights; the other
would not. Their different methods, as well as the ensu-
ing public debate between them over their philosophies,
helped determine whether the telescope would make the
transition from a purely qualitative instrument—one
that showed only what was out there, astonishing as that
may be—to a quantitative tool as well.

A century earlier the king of Denmark had rewarded
the astronomer Tycho Brahe with a small island and a
ton of gold to finance the finest observatory ever. Over
several decades, at the retreat he called Uraniborg (a
combination of Greek and German meaning "city of
heaven"), Tycho had amassed the most sensitive instruments in astronomical history, developed a new
approach to observations that monitored and measured
celestial bodies throughout their cycles, rather than only
on such momentous occasions as eclipses and conjunc-
tions, and repeated those observations over years and
decades, sometimes as many as seven times. He also
possessed superior eyesight. As a result, his observations had improved on ancient observations by as much
as a factor of fifty. Now a new generation had set itself
the task of improving on *that*.

"We have heard that the celebrated Johannes
Hevelius has indeed undertaken the restitution of the

fixed stars," the British astronomer John Flamsteed
wrote in July 1673, "yet seeing that he is reputed to use
sights without glasses [lenses], it is doubtful if we shall
obtain from him much more correct places than Tycho
left us, except where he went very much astray."

The words wounded Hevelius. Not only was one of
the leading astronomers of the younger generation
questioning his methods, but he was doing so in the
pages of the *Philosophical Transactions,* the official
organ of London's Royal Society, the most prestigious
and influential collection of scholars of the day.
Hevelius responded in writing and at length, defending
the accuracy of his measurements and offering to sub-
stantiate his high opinion of them. Before the decade
was out, the Royal Society sent the young Edmond
Halley, the so-called southern Tycho who had cataloged
the heavens visible from the Southern Hemisphere
(using telescopic sights), to visit the elderly Hevelius in
his observatory in Danzig. "I assure you I was surpriz'd
to see so near an agreement in them, and had I not seen,
I could scarce have credited the Relation of any," Halley
wrote to Flamsteed in June 1679, after comparing his
own observations with those of Hevelius. "I dare no
more doubt of his Veracitye."

In a way the battle was generational—between one
era of astronomer and the next. Eventually the disagree-
ment degenerated into little more than name-calling,
with Halley dismissing Hevelius as "a peevish old gen-
tleman, who would not have it believed that it is possible to do better than he has done." Still, what was at
work here was a factor far more subtle than age or stub-
bornness.

Hevelius was no novice. Over the preceding four decades he had established in Danzig what was for a time the world's leading astronomical observatory. He had started in a small upper room, added a roofed tower, then constructed a fifteen-hundred-square-foot platform supporting two observation houses, one of which could revolve. His *Selenographia,* published in 1647, was the first lunar atlas. It set the standard for the artistic representation of celestial observations for decades to come; provided a highly detailed, lavishly illustrated guide to lenses, lathes, glass quality, lens combinations, tube construction, and apertures; and served as the essential introduction for amateurs and professionals alike in the areas of telescope construction, housing, and mounting. The problem that Hevelius presented to Flamsteed, Halley, and the others, then, was this: He was one of the most distinguished observers in the relatively brief history of the telescope, one of the most experienced and most experimental, and he *still* preferred the naked eye when it came to calibrations.

For in his long experience Hevelius had learned the untrustworthiness of the new instrument. In 1647, in his *Selenographia,* he had disproved the existence of numerous new satellites around Jupiter, Mars, and Saturn, recently "discovered" by the same Capuchin monk who had introduced the terrestrial telescope and "confirmed" by Francesco Fontana. But then, in the same volume, Hevelius had made his *own* blunders, claiming to have seen finite disks for the fixed stars and illustrating the phases of Mercury in reverse order— both because of phantom images in his telescope. In addition, during the following decade he published his conclusion, supported by illustrations, that Saturn was in the shape of an egg bracketed by two attached arcs. When Huygens published his own ring hypothesis for Saturn later in the decade, Hevelius wrote to a colleague: "Does Huygens perhaps suppose that I and others are not able to discern what is elliptical or spherical? No, by Hercules!"

For Hevelius, experiences such as these provided all the proof he needed that as he began the most delicate and ambitious observations of his life, he shouldn't trust the telescope. For his younger contemporaries, however, these same experiences were the kind that demonstrated the need for results that were independently verifiable. All these observers, Hevelius included, belonged to the age of the telescope, the first generations to do so. They all understood that the telescope had inaugurated not only a new kind of investigation into the heavens but a new kind of investigation, period—a New Philosophy that placed more faith in individual observation than in ancient authority. As Hevelius himself had inscribed on the title page of one of his books, "Not by words but by deeds." Where the two generations differed was in how best to perform those deeds. "I prefer the unaided eye," Hevelius had also inscribed on a title page, but his rivals preferred measurements that posterity could attempt to duplicate and thereby judge.

Aristotle had trusted in the evidence of his senses. Galileo had trusted in the evidence of the first instrument to extend one of the human senses. Now a new generation of astronomers was making a further distinction. Even if the eyesight of a Hevelius was to prove more piercing than even the great Tycho Brahe's, it was still only human. What astronomy needed was a standard that wasn't. The new generation of astronomers trusted in evidence from the telescope, but they trusted in it even more when it didn't depend on the interpretation of the observer; when it was answerable to the higher power not of ancient authority, or even of God, but of Nature; when it was quantifiable, measurable, replicable, absolute—when it was, in a word, mechanical.

The idea of a mechanical universe was nothing new. It was inherent in the *nuova arte* interpretation of flat space in three dimensions; it was implicit in Galileo's declaration that the Book of Nature is written in geometrical characters. "My aim," Kepler had written, "is to show that the heavenly machine is not a kind of divine, live being, but a kind of clockwork," while the English chemist Robert Boyle said the natural world was, "as it were, a great piece of clock-work." In part the clockwork comparison made sense because the complex inner workings of timepieces represented the greatest degree of mechanical precision known to man.

But in part it made sense because time, after all, *came* from the sky. Time was how man measured the rising of the Sun, the phases of the Moon, the return of the equinox: day, month, year. Whether the celestial sphere was moving around the Earth, or the Earth was spinning through space on its own axis, the effect was the same: A day was a day because that was how long it took the heavens to appear to complete one revolution.

But how long was a period shorter than a day? The ancients had divided the circle that the spinning stars describe on the sky every day into 360 parts, or degrees (360 being a number easily divisible by many other whole numbers). Divide that circle into 24 hours, and the result is a movement of 15 degrees per hour. Do some further division, and the result is a movement of 15 arc minutes per minute of time and a movement of 15 arc seconds per second of time. Now attach a telescopic sight to instruments that measure angles on the celestial vault in arc minutes and arc seconds, train that sight on a celestial object, and track that object to the sound of a ticking timepiece—possible after Huygens, in 1656, had refined the pendulum weight—and the idea of a clockwork universe begins to take on a whole new and literal meaning.

Nine hours, fifty-six minutes: This was the period of Jupiter's rotation, according to Cassini's observations of the markings on the planet's surface. For Mars, he determined a period of twenty-four hours, forty minutes. He also succeeded in compiling tables that predicted the movements of the moons of Jupiter, a task that had defeated even Galileo.

From those tables, the Danish astronomer Ole Römer noted that eclipses of the various moons of Jupiter seemed to differ depending on where Jupiter and Earth were in relation to each other in their orbits around the Sun. When Jupiter and Earth were farther apart, the eclipses and transits seemed to occur later than predicted; when closer together, earlier than predicted. Could these discrepancies be due to light traveling not instantaneously, as most astronomers believed it to do, but "gradually"? If so, could timing the intervals help calculate a figure for the speed of light?

In September 1676, Römer announced to members of the Académie des Sciences in Paris that the eclipse of the innermost satellite of Jupiter expected at 5:25:45 A.M. on November 9 would occur ten minutes late. It did: 5:35:45 A.M., exactly. From this observation he calculated that the speed of light was such that it would take twenty-two minutes to cross the full diameter of Earth's orbit around the Sun—or, on the basis of the best estimates of the day, 140,000 miles per second.*

"It is so exceeding swift," wrote the English astronomer Robert Hooke, on hearing the news, "that 'tis beyond Imagination; for so far he thinks undubitable, that it moves a Space equal to the Diameter of the Earth, or near 8000 Miles, in less than one single Second of the time, which is in as short time as one can well pronounce 1, 2, 3, 4: And if so, why it may not be as well instantaneous I know no reason." In other words, some distinctions might not be worth making. At the time Hooke was the most severe critic of Johannes Hevelius's refusal to use telescopic sights in assembling a catalog of the stars, yet like Hevelius he couldn't imagine why a standard of precision even subtler than the present one might be possible or important.

By 1675 astronomers had reached consensus on the apparent sizes of planets, the diameters observable from Earth. These results, however, were only relative; they

* Modern value: 186,282 miles per second.

reflected the relationship of the planets to one another, but not their absolute sizes. For that, astronomers would need to know the distances of the planets from Earth at various points in their orbits, and for *that,* they would need a standard unit of measurement.

And that measurement would have to involve the one component that all the elements have in common: the center around which they revolve. In the old Kosmos—the Aristotelian, Ptolemaic, geocentric universe—the distance from man to the center of all existence had been the Earth's radius. But in the new cosmos—the Copernican, Galilean, heliocentric one—the standard unit would be the distance to a different center: from the Earth to the Sun.

After inventing the telescopic sight, Jean Picard walked around the neighborhoods of Paris, using the new instrument to conduct more accurate land surveys. As he did so, he arrived at a new measurement of the length of a terrestrial arc at that latitude, and from that he was able to arrive at a new approximation of the diameter of the Earth. These determinations in turn allowed Jean Richer, during an expedition to the island of Cayenne in South America, to compare measurements of Mars near the Equator with those from Paris, which in turn led to a new approximation of the distance to Mars, which in turn led to a new approximation of the distance from the Earth to the Sun, or the astronomical unit: 82 to 87 million miles.*

In a way, this was literally a difference of degrees.

* Modern value: 92,955,807 miles.

Tycho had been able to measure the position of a star within 3 arc minutes of accuracy, by far the best result up to his era; Flamsteed, however, could achieve measurements within 10 arc *seconds*. With a screw micrometer, Flamsteed could measure the position of Mars with respect to the stars within 1 arc second—$\frac{1}{3600}$ of a degree, or the distance that the planet would appear to travel in its daily circle on the celestial vault over the course of $\frac{1}{15}$ of a second. Put enough of these degrees, minutes, and seconds together, and they *did* add up to a difference in kind.

It wasn't only what the telescope had helped reveal qualitatively—the celestial bodies visible in a variety that until very recently would have been beyond imagination. It wasn't even what the telescope had helped render quantitatively—the dimensions and distances that approximated the way the universe actually works. Instead what had changed astronomy at least as much as these observations was faith in the telescope itself. When two great astronomers of the 1670s set out to survey the sky, their intentions might have been identical—to improve on Tycho—but their philosophies were not. Maybe Hevelius was right, and his naked eye could match the telescope measurement for measurement, for now. But his insistence made sense only as long as the naked eye remained the standard frame of reference, which would be only as long as the telescope stayed the same. Once it improved, Flamsteed's observations became the foundation for the future.

The three volumes of Hevelius's catalog of the heavens were published, in 1689, within a couple of years of his death. Flamsteed's, based on twenty thousand observations of the fixed stars between 1676 and 1689 using a sextant fitted with telescopic sights, followed more than thirty years later. It was an indication of just how spectacularly Hevelius had miscalculated the usefulness of his data that even during this three-decade interim, the members of the Royal Society didn't resort to Hevelius's published observations, preferring instead to pester Flamsteed about getting his own results into print. Flamsteed, however, was notoriously perfectionistic; when Halley published a volume of Flamsteed's results without his permission, in 1712, Flamsteed rounded up most of the copies and burned them. It wasn't until 1725, six years after his death, that the three-volume *Historia Coelestis Britannica* finally saw print, a catalog of three thousand stars, far and away the most extensive survey of the skies to date.

The universe they left behind—Flamsteed, Hevelius, Halley, Huygens, Cassini, and all the other astronomers who struggled with the faulty lenses, clumsy lengths, and frustrating optics of the second half of the seventeenth century—was not the universe they found. They had succeeded in determining, even if only approximately, the sizes of the planets, their distances to the Sun, and how far were the stars. At the start of the century the prevailing scale of the universe was the same one that Ptolemy had set out: the Sun as 1,200 Earth radii distant, or 5 million miles, and the celestial vault of fixed stars as 20,000 Earth radii distant, or 80 million miles—enormous distances, to be sure.

By the end of the century, however, the distance from

the Earth to the Sun alone was greater than the radius of the entire Ptolemaic Kosmos; the radius of the orbit of Saturn, and therefore of the solar system, was some 800 million miles; and the very fact that astronomers had begun to think in terms of a "solar system" was itself an indication of the distance they were beginning to put between themselves and the stars. The new universe that the astronomers at the turn of the eighteenth century found themselves inhabiting was one where the fundamental division was no longer between what was terrestrial and what was celestial, but one where the Earth's closest kinship was with its fellow planets, where the distance to the stars demanded a previously unthinkable new scale in the thousands of millions of miles, where, if the evidence of the astronomical telescope could be believed—and with every new measurement there was more reason to think it could—up was down, and down was up.

✳

CHAPTER FOUR

Profundity

William Herschel was going home, but he was thinking of the stars.

Actually he was leaving home too. He had spent several days visiting family in his birthplace of Hanover, Germany, and now he was bringing his sister, Caroline, back with him to England, where he'd settled fifteen years earlier. During his years abroad he had established himself as a successful composer and performer, and he hoped that he might be able to provide his sister with opportunities for training and advancement as a singer. He taught as many private students as his schedule allowed, often more than thirty-five lessons a week; he composed concertos, anthems, psalms, and symphonies; he performed in concert on the violin, oboe, harpsichord, and organ; and his position as church organist with the socially prominent Octagon Chapel in the fashionable resort town of Bath placed him squarely in circles of influence and refinement. But on the open seat of a mailcoach as it crossed the windswept roads of

Holland, halfway between homes, what William
Herschel talked to his sister about instead were the constellations.

His fascination with the stars had grown out of his
interest in music. Several years earlier, in order to edu-
cate himself about the mathematical aspects of music,
he had sought out *Harmonics,* by the Cambridge
astronomer Robert Smith. That book led him to *A
Compleat System of Opticks,* by the same author, and
from there to other books on optical principles, especially as they related to the emerging study of astronomy. "When I read of the many charming discoveries
that had been made by means of the telescope," he later
recalled, "I was so delighted with the subject that I
wished to see the heavens and planets with my own eyes
thro' one of those instruments."

Herschel kept a journal, but only of the most per-
functory sort. Weeks might pass between entries, or
months; at one point two years went by before he found
something he considered worth noting. Even when he
did make an entry, its spareness could speak volumes
about the burden of obligation he must have felt, either
toward fulfilling his official duties or toward keeping a
journal. "Concert, Linley," went one, in its entirety. And
another: "Played the Organ. Took the Sacrament."
Shortly after his visit to Hanover and the ride under the
open night skies of Holland, however, his journal began
to exhibit a lighter tone—very nearly, for Herschel, an
expansiveness.

The final two entries for 1772 record his return from
Hanover with his sister and his resumption of duties:

Aug. 27. Arrived at Bath.

Sept. 1. Began again to teach my resident scholars.

Then come the first entries for 1773:

April 7. Oratorio at Bristol.

April 8. Concert Spirituale.

April 19. Bought a quadrant and Emerson's *Trigonometry.*

May 10. Signor Farinelli's Concert.

May 10. Bought a book of astronomy and one of astronomical tables.

May 24. Bought an object glass of 10 feet focal length.

June 1. Bought many eye glasses, and tin tubes made.

June 7. Glasses paid for and the use of a small reflector paid for.

June 14. Boxes for glasses paid for. The hire of a 2 feet reflecting telescope for 3 months paid for.

June 21 to Aug. 23. Many glasses, tubes.

Sept. 15. Hired a 2 feet reflector.

Sept. 22. Bought tools for making a reflector. Had a metal cast.

Oct. 2. Bought a 20 feet object glass and nine eye-glasses, etc. Emerson's Optics. Attended private scholars as usual.

Nov. 8. Attended 40 scholars this week. Public business as usual.

Nov. 15. Attended 46 private scholars; nearly 8
    per day.

It was the longest year's worth of journal entries
since his arrival in Bath, and it barely began to suggest
the extremity of the transformation Herschel's life was
undergoing. Astronomy now consumed him. He bought
optical tools and workshop equipment from a neighbor
who built telescopes as a hobby, and he began devoting
all his spare time to his newfound passion. Between acts
at concerts he raced outside to study the skies. Every
night he went to bed reading books about optics and
astronomy, and "his first thoughts on rising," his sister
wrote in her own (far more copious) journal, "were how
to obtain instruments for viewing those objects himself
of which he had been reading." At times, recalled
Caroline, whose nascent singing career suffered in
direct proportion to her brother's astronomical studies,
"by way of keeping him alife I was even obliged to feed
him by putting the Vitals by bits into his mouth;—this
was once the case when at the finishing of a 7-foot mirror he had not left his hands from it for 16 hours
together."

Similarly, Herschel never missed a single hour of
clear weather for observation. Even when he changed
residence, in 1786, his sister noted that "the last night at
Clay Hall was spent in Sweeping [the sky] till daylight,
and the next the Telescope stud ready for observation at
Slough." Herschel always observed in the open because
his telescopes performed well only when they had
reached the same temperature as the night air, which,

during the English winter, often fell well below freezing, even down to the single digits Fahrenheit. His sole concession to the elements was to wear extra layers of clothing. He would rub himself with a raw onion to combat the ague, while his breath crystallized on the side of the telescope tube, his feet sank into the mud, the ink congealed in its well, and, on one occasion, the mirror itself snapped in half with a crack like a rifle report. As a German astronomer wrote after witnessing Herschel at work, "He has an excellent constitution and thinks about nothing else in the world but the celestial bodies." In this single-minded dedication and fortitude he was matched by his sister, Caroline. When she fell one New Year's Eve in a foot of melting snow and snagged her right leg above the knee on an iron hook, she took care to note, "I had, however, the comfort to know that my Brother was no loser through the accident, for the remainder of the night was cloudy."

At that time astronomy was not an uncommon way for a man of education to occupy himself in his leisure hours. Over the preceding century the telescope had become the foremost symbol of learning, an emissary of civilization's best hopes for the powers of investigation and imagination. Even so, Herschel's technical talents and intuitive flights distinguished themselves. When he couldn't find telescopes that met his exacting standards, he learned to construct his own, and he applied them toward methodical surveys of the sky that he hoped might help him map the stars.

It was on March 13, 1781, deep in the second and final year of his second review of the stars, a catalog of every star in the sky down to a certain level of luminosity, that William Herschel encountered a celestial object that shouldn't have been there. He knew it wasn't a star; it clearly was in the shape of a disk. He switched magnifiers, and the disk grew proportionately, while the surrounding stars, as expected, did not. He switched magnifiers again, and again the disk grew. He returned to the object the following night and calculated from its shift in position that it lay within the solar system. He wrote up his observations under the title "Account of a Comet" and submitted it to the Royal Society in London. He soon heard back that other astronomers had followed his directions and sought out the object, and that after making their own calculations, they had concluded that Herschel was right in assuming it wasn't a star but incorrect in calling it a comet. What he'd found instead was a new *planet.*

In the history of the world, there had always been one reliable roster of wanderers: Mercury, Venus, Mars, Jupiter, Saturn, as well as the Sun and Moon. In 1610 Galileo had overturned many ancient assumptions about the constitution of the heavens, even adding new moons to the night sky, but (with the possible exception of the addition of Earth itself) he hadn't changed the number of planets. Even the lunar population had remained constant for nearly a hundred years, since Cassini's discovery of a fourth satellite of Saturn in 1684. But now an amateur had undermined not only ancient authority but current authority as well and in the process had doubled the diameter of the known solar system.

To the words "solar system," itself a notion of

extraordinarily recent and profound vintage, the cautious could now add the qualifier "known," and not without some satisfaction. In 1781, in recognition of the discovery of the first planet since the dawn of astronomy, the Royal Society awarded Herschel its highest honor, the Copley Medal, and at the presentation the president of the organization asked, "Who can say but your new star, which exceeds Saturn in its distance from the sun, may exceed him as much in magnificence of attendance? Who knows what new rings, new satellites, or what other nameless and numberless phenomena remain behind, waiting to reward future industry and improvement?"

What a difference a century makes. At first the outpouring of evidence available through the telescope had destroyed the old hierarchy of the universe, while replacing it only with a vague unruliness, an actual lack of rules. For some this uncertainty was too much. "And new Philosophy calls all in doubt," the English cleric and poet John Donne had written in 1611, in large part in response to Galileo's *Sidereus Nuncius,*

> The Element of fire is quite put out;
> The Sun is lost, and th' earth, and no mans wit
> Can well direct him where to looke for it.

By the time of Herschel's discovery of the planet that would become known as Uranus, however, the telescope had rewarded astronomers' faith in the instrument itself, as well as in the New Philosophy it both represented and furthered. It had helped create a new order for the universe, one that could not only accommodate the addition of a new planet but warmly welcome any others to come.

Seeing was one thing, believing another, and as long as astronomy through the telescope had remained a purely subjective discipline—as long as it was observational but not positional, qualitative but not quantitative—its findings were open to debate and doubt. Even the introduction of the micrometer and the telescopic sight hadn't freed astronomy of controversy, as Johannes Hevelius had demonstrated. But precision measuring instruments had allowed astronomers to begin to establish standards that were objective, had given them at least a chance to achieve some degree of certainty.

Now they could believe: The universe operates according to rules, and it is possible to learn those rules. This was not the belief of the ancients, whose "world on paper," as Galileo and Kepler had dismissively called it, was the product of mathematical acrobatics sufficient only to save the appearances. This *was* the belief of Galileo, but his observations had produced a new set of appearances, not the math to match. Only when astronomers had begun to derive precise quantitative data in the latter half of the seventeenth century—the distances and dimensions of the solar system—could they begin to answer the question: If you drop a rock on a speeding, spinning planet, why *does* it land here, not there?

In 1609 Johannes Kepler had published a book called *Astronomia nova* (or *New Astronomy*), using the wealth of data that his mentor Tycho Brahe had accumulated at Uraniborg. Adopting *nova* in a title of anything was hardly novel in those days, but in this case it was entirely apt, for Kepler, after hundreds of pages and several years of calculations, had refined Tycho's observations into a pair of laws that overthrew two fundamental notions of Aristotelian physics—that celestial objects move in circles, and that those movements are uniform. First, Kepler wrote, the orbit of each planet is an ellipse. Second, while traveling along that ellipse, the planet slows down as it moves away from the Sun and speeds up as it nears the Sun. Ten years later Kepler added a third law: The farther a planet's average distance is from the Sun, the longer it takes to orbit around the Sun; the nearer, the shorter—specifically, that the square of the time it takes a planet to complete one orbit around the Sun is proportional to the cube of the average distance of the planet from the Sun. For the first time in history, astronomers had at least the makings of a mathematics that actually accorded with observations.

In these laws Kepler had given mathematical expression to Tycho's observations, the most precise measurements of the pretelescopic era. Any fuller mathematical expression of the movement of heavenly bodies, however, would need to await even more precise measurements. For this reason Kepler's work didn't become standard reading material for astronomers until the mid-1660s. Only after the introduction of the astronomical telescope and its quantitative adjuncts, the micrometer and the telescopic sight, could astronomers begin making the kinds of observations with sufficient precision to render Kepler's work accessible, then indispensable.

In his *Philosophiae naturalis principia mathematica* (or *Mathematical Principles of Natural Philosophy*), in

1687, Isaac Newton effectively overthrew Aristotelian physics once and for all. When he thought about gravity, he wondered how far it extended. If it reached to the top of an apple tree, could it also reach to the Moon? He applied Kepler's third law to this idea, added the terrestrial and celestial measurements that Jean Picard and Jean Richer had made in Paris and Cayenne respectively, and arrived at a formula of his own that in effect condensed all three of Kepler's laws and potentially pretty much everything else in the universe: Every object attracts every other object with a force that is inversely proportional to the square of the distance.

Forget the math for a moment. *Every object attracts every other object:* This was a leap that any artist or astronomer could admire and one that wouldn't have been possible without their earlier efforts. In the preface to the second edition of the *Principia,* the English astronomer Roger Cotes framed this principle in an analogy that would have appealed to any reader of the day: "For who doubts, if gravity be the cause of the descent of a stone in Europe, but it is also the cause of the same descent in America?" If what was true in the Old World was true for the New, then what was true in this world was true for the next. The reason a rock landed here, not there, was the same reason the Moon followed the Earth, and the Earth the Sun: gravity.

Not that Newton could say what gravity *was.* "Action at a distance," he called it, inciting the doubts and ridicule of peers who wanted to know what this action was, how it covered the distance, and what medium, if any, it had to pass through along the way. "I feign no hypotheses," he answered them. "Gravity must be caused by an agent acting constantly according to certain laws; but whether this agent be material or immaterial, I have left to ye consideration of my readers." All he knew for sure was that he had satisfied the math. Whatever gravity was, it must work, because *he had the math to prove it.*

This, then, was the Book of Nature written in geometrical terms that Galileo had advocated and had himself done much to transcribe; this was celestial mechanics rendered as mathematical abstraction; this was the end of ancient authority and the ascension of a modern one.

This was the way the universe works.

"Eppur si muove," Galileo supposedly (and, alas, no doubt apocryphally) said after recanting his faith in Copernicanism before the Inquisition: *And yet it moves.* But the Earth, it turned out, didn't just move. It moved in ways equally predictable and verifiable. And the more astronomers looked, the more it moved; and the more it moved, the more it confirmed their belief not only that the universe operates according to rules, and not only that it was possible to learn those rules, but that they, for the most part, had done just that.

The experiment was a success: The New Philosophy worked: A new tool had procured evidence from the natural world that fostered a new understanding of the universe. "On all sides attentive eyes are fixed on Nature," the French economist and statesman Anne Robert Jacques Turgot wrote, in 1750, in *Discourse at the Sorbonne.*

Slight chances turned to profit bring forth discoveries. The son of an artisan in Zeeland,
while amusing himself, brings together two
convex glasses in a tube, and the limits of our
senses are removed. In Italy the eyes of
Galileo have discovered a new celestial world.
Now Kepler, while seeking in the stars the
numbers of Pythagoras, has found the two
famous laws of the course of the planets
which will become one day, in the hands of
Newton, the key to the universe. . . . At last all
clouds are dissipated. What a glorious light is
cast on all sides! What a crowd of great men
on all paths of knowledge! What perfection of
human reason!

It was the Age of Reason, the Age of Confidence, the
Age of Criticism, and the Philosophical Century. It was
the Age of Enlightenment, an epoch that took its name
and its organizing principle from the new hierarchy of
the heavens: a central Sun, enlightening all. The idea of
a clockwork, cause-and-effect universe might not have
been new, but it gained currency when Newton claimed
in the *Principia* that "the Copernican system of the
planets stands revealed as a vast machine working under
mechanical laws here understood and explained for the
first time." In the century to come, astronomers proved
again and again that they could take the numbers available from measurements made through the telescope,
plug them into Newton's formula, and both predict and
account for the movements of heavenly bodies with

equal accuracy. Even seeming refutations to a law of gravity—irregularities in the orbits of Jupiter and Saturn, the return of a comet in 1758 later than Edmond Halley had predicted (the comet got his name, anyway)—turned out, on closer inspection and after further calculations, only to confirm it. "Such has been the fate of [Newton's] brilliant discovery," wrote the French mathematician Pierre Simon Laplace, "that each difficulty which has arisen has become for it a new subject of triumph, a circumstance which is the surest characteristic of the true system of nature."

In his *Traité de mécanique céleste* (or *Treatise on Celestial Mechanics*), published in several volumes beginning in 1799, Laplace took Newton's theory to a logical extreme. He set himself the task of explaining the motions of every known celestial object in the solar system according to the law of gravitation—moons, planets, Sun, more than thirty objects in all, each exerting its own mathematically calculable effect on every other object, and vice versa, in a dizzying dance of mutual attractions. Like a curious child taking apart a timepiece, Laplace couldn't afford to leave out the least spring or screw when putting his clockwork universe back together. Somehow, he succeeded:

> It is very remarkable that an astronomer, without leaving his observatory, by merely comparing his observations with analysis, may be enabled to determine with accuracy the magnitude and flattening of the earth, and its distance from the sun and moon, elements the

knowledge of which has been the fruit of long and troublesome voyages in both hemispheres. The agreement between the results of the two methods is one of the most striking proofs of universal gravitation.

Precision begat progress, and progress begat perfection. The ability to make finer measurements led to an expectation that greater knowledge would accrue to each passing generation. The acquisition of that knowledge in turn nurtured a belief in, eventually, absolute comprehension. As Laplace hypothesized, "an intelligence knowing, at a given instance of time, all forces acting in nature, as well as the momentary position of all things of which the universe consists, would be able to comprehend the motions of the largest bodies of the world . . . [and] nothing would be uncertain, both past and future would be present." Like the artists of the *nuova arte* several centuries earlier, the astronomers of the modern era had parted a curtain and discovered a universe of geometry. In place of the old hierarchy of the universe stood not the anarchy that anti-Copernicans had initially feared but its opposite: a few simple rules that anticipated and explained the movement of each piece in the great clockwork mechanism of the cosmos, a single law that governed all physical behavior from here to the end of the universe. Which left only one question: Just where *was* the end of the universe, anyway?

If William Herschel didn't quite set out to answer this question, it nonetheless was the issue that was beginning to inform all the astronomical work of the day.

Galileo had inherited a Kosmos that divided between the terrestrial and the celestial, between the Earth and everything else, and he had found it confining. In the same way, Herschel now inherited a universe that divided neatly in two—between the newly accessible planets of this solar system and the still-unreachable recesses of the stars—and he assigned himself the task of parting this perhaps final curtain.

"Hitherto the sidereal heavens have, not inadequately for the purpose designed, been represented by the concave surface of a sphere," Herschel once wrote, "in the centre of which the eye of an observer might be supposed to be placed." Astronomers at this time knew better, but they still couldn't help conceiving of the celestial vault as a two-dimensional surface, a grid on which to plot the positions of stars. Herschel, however, sought to probe the third dimension that had to be there. "The construction of the heavens," Herschel wrote, "in which the real place of every celestial object is to be determined, can only be delineated with precision, when we have the situation of each heavenly body assigned in three dimensions, which in the case of the visible universe may be called length, breadth, and depth; or longitude, latitude, and Profundity."

Long before Herschel started observing, progress in stellar astronomy had slowed to a virtual standstill. Part of the reason, as it often was at such moments, was the mechanics of the instrument. Just as the narrow field of view in the Galilean model of the telescope had set the limits for an earlier generation of astronomers, restricting them to the qualitative work of observing and recording new phenomena, so the long refractor turned out to have come with its own inherent limitation. Despite the many advances in the telescope's conversion from a crude observational tube into a precision quantitative instrument, even the most delicate examples of the day hadn't been able to determine the one seemingly fundamental measurement in the investigation of the stellar regions: the distance, or parallax, of a single star.

Parallax is a form of triangulation similar to how our eyes judge distances. Hold up a finger and close one eye; now open that eye and close the other; repeat; repeat again. The way the finger appears to be jumping back and forth is an exaggeration of how a star should appear to be jumping back and forth in a Sun-centered universe, as the Earth moves in its orbit. In fact, the absence of parallax had always presented one of the conundrums of Copernicanism. As long as the Earth stood at the center of the universe, unmoving, the fact that the stars were "fixed" on the celestial vault made sense. But if the Earth moved around the Sun—if the Earth did wheel from one side of the Sun to the other every six months, and then back again—then the fixed stars should appear to be moving as well, however slightly. But they didn't. In order for Copernicanism to hold, the distance to the stars had to be enormous enough to render the Earth's movement minuscule, insignificant. In comparison with the distance to the stars, the diameter of the Earth's orbit around the Sun would have to be *as nothing*—and after the first determination of the Earth-Sun astronomical unit, in the

1670s, the diameter that was as nothing was at least 170 million miles.*

To be so distant yet so bright, stars had to be of comparable size and luminosity to our Sun. By the late seventeenth century astronomers were beginning to reach a consensus that that was what stars were: so many suns, most perhaps hosts to their own systems of planets. This hypothesis gained considerable support in 1718, when Edmond Halley announced that he had detected not parallax but proper motion, not a shift caused by the Earth's movement in relation to a star, but a change in a star's position on the celestial vault attributable to the motion of the star itself. The movements he found were detectable only by comparing present positions with tables of observations from two thousand years earlier, and they were so slight that Halley could find only three examples. Still, motions they were, meaning that at least these three stars—and presumably many more, if not all—were "fixed" in name only.

Ten years later another announcement helped place stellar distances in an even more daunting perspective. The English astronomer James Bradley revealed that not only had he been unsuccessful in his search for a movement in a star caused by the motion of the Earth— the elusive parallax—but he'd determined that he would have been able to detect one if it had been greater than 1 arc second, or the width of a coin at a distance of several kilometers. From this finding he calculated that the distance to the nearest star must be at least 400,000

---

* Modern value: 186 million miles.

astronomical units, or 36 trillion (36 million million) miles.

The great divide between the solar system and the stars, already unimaginably vast, was now also unbridgeable. Astronomers had literally gone to great lengths to overcome the optical defects of refracting telescopes, and they had been rewarded with measurements several magnitudes more accurate than those available to pretelescopic observers. Those measurements in turn inspired theoretical astronomy of the kind that Pierre Simon Laplace practiced, in which mathematicians attempted to navigate the solar system through calculations based on Newtonian gravitation. But in the absence of either a way to overcome the optical problems of the refractor or a new telescopic development altogether, practical astronomy wasn't going to be making significant advances in the distant stellar realm—if indeed there were even any advances to make.

In fact, though, a bold, breakthrough innovation *had* arrived decades earlier, in the form of the reflecting telescope—a telescope that employed not lenses to bend, or refract, light but a mirror to reflect it. In the 1660s and early 1670s several mathematicians (including Isaac Newton) had realized that a mirror in a telescope would perform the same function as a refracting lens but without chromatic aberration, the color fringes that plagued refractors. In January 1721 James Bradley presented to the Royal Society a reflector that through trial and error he had succeeded in ridding of spherical aberration as well, by grinding the mirror into a paraboloid surface, the only shape that would bring all the rays of light to the same focus. His 6-foot reflector proved to be far more manageable than the 123-foot refractor the Royal Society was then using, and it also performed with comparable magnification and definition. Over the coming decades, as grinding techniques improved, the reflector became an extremely popular alternative to the long refractor, especially among amateurs. Among professionals in need of precision measurements using a micrometer, however, the long refractor remained the telescope of choice.

Herschel himself started with refractors. With the kind of naïveté available only to a true amateur, he had decided from his reading that what he wanted to contribute to the world of astronomy was a measurement of stellar parallax. First he used a 4-foot refractor; then came a 12-foot, a 15-foot, and finally a 30-foot. Their slow maneuverability frustrated him, however, so he rented a reflector only 2 feet long. He found that a mirror could deliver not only comparable magnification with more convenience, but an additional advantage that, he came to realize as he graduated to larger sizes, could be of even greater value for an astronomer primarily interested in the stars: more light.

The telescope in fact had never been exclusively a magnifying instrument. Its magical power to bring distant objects seemingly near enough to touch was only its most obvious attribute, and the temptation to play with the ratio of focal lengths between the objective and the eyepiece and, in so doing, make the image bigger and bigger and bigger—to exploit the magic for all it was worth—had been understandably irresistible.

The telescope, however, was no less a light-gathering instrument. A lens or mirror can easily gather more light than the human retina; how much light depends on the surface area of the opening, or aperture—specifically, the square of the diameter. Double the diameter of the aperture, and its light-gathering capacity increases four-fold; triple it, and the capacity goes up ninefold. At the same time the brightness of the object under observation depends on the square of the *distance*. Double the distance of a light source, and its brightness *decreases* fourfold; triple it, and the brightness drops off ninefold. The implication was clear: Double the diameter of the aperture, and you double the distance the mirror can see. Herschel understood that for the purpose of investigating the starry depths the major advantage of a reflecting telescope wasn't in greater magnification, but in gathering more light—wasn't in seeing more detail, but in seeing *farther*.

The emphasis on magnification in astronomy, it turned out, had acted somewhat as a self-fulfilling prophecy. As with the Galilean telescope and the belief that Galileo had found at once all the new phenomena worth finding, the limitation in the long Keplerian telescope wasn't merely mechanical—wasn't only in the degree of specificity with which the telescope could measure angles in an attempt to ascertain parallax. It was in the assumption that the stars were too far away to explore. Magnification, of course, makes sense only when the phenomenon under observation is one that would benefit from greater detail, such as a planet, a satellite of a planet, a crater on the Moon, a sunspot—

something nearby in astronomical terms, something within the solar system. But a star doesn't benefit from magnification; magnify a pinpoint of light, and it remains, resolutely, a pinpoint of light. So astronomers didn't bother with the pinpoints of light, and every observation, every measurement that rendered the fellow members of the solar system closer to Earth and cast the stars farther away, only reinforced this prejudice, until by the time Herschel started observing, the stars were little more than an astronomical afterthought.

"When, in the course of time, I took up astronomy, I determined to accept nothing on faith," Herschel once wrote, "but to see with my own eyes everything which others had seen before me." A good thing too: At the time the number of stars in the *British Catalogue* was around 3,000, and James Ferguson's *Astronomy,* one of the books with which Herschel retired to bed on a regular basis, devoted only one of its twenty-two chapters to anything other than the solar system. The one chapter it did devote to stars noted that their number "is much less than generally imagined" and concluded: "There is a remarkable tract round the Heavens called the Milky Way from its peculiar whiteness, which was formerly thought to be owing to a vast number of very small stars therein; but the telescope shows it to be quite otherwise; and therefore its whiteness must be owing to some other cause." By contrast, Herschel once estimated that during one forty-one-minute period 258,000 stars passed across his mirror.

In terms of his observing program, then, his discovery of Uranus was an anomaly, but a fitting one. A new planet approximately twice the distance from the Sun as

Saturn provided a significant first step, a link literal and metaphorical, between the ancient roster of wandering "stars," or planets, and the realm of the "fixed" stars, or the celestial vault—between the newly christened solar system and everything else. While astronomers (and the rest of the world) at first thought they were welcoming the news of a delightful discovery in the otherwise familiar neighborhood of the solar system, they soon found that the greater significance lay in the new technology an amateur astronomer was wielding on a new frontier—or, rather, an existing technology put to such new purposes that it forced a rethinking of both the technology and the frontier.

Herschel's initial report to the Royal Society about a comet inspired open speculation about whether this amateur correspondent was "fit for Bedlam," not because his discovery turned out to be a planet but because of the means by which he'd discovered it. A friend wrote Herschel from the Royal Society: "What! say your opposers, opticians think it no small matter if they sell a telescope which will magnify 60 or 100 times, and here comes one who pretends to have made some which will magnify above 6000 times! is this credible?"

"From the contents of your letter," Herschel replied, "I begin to have a much better opinion of my own observations than I had before. I thought what I have seen had been within the reach of many a good telescope." In the months to come, he dutifully packed up his equipment, carted it to London, and demonstrated it for the astronomer royal, Nevil Maskelyne. "Among opticians & Astronomers nothing is now talked of but *what they call* my great discoveries," he wrote to Caroline from London. "Alas! this shews how far they are behind when such trifles as I have seen and done are called *great*. . . . Dr. Maskelyne has already ordered a model to be taken from mine and a stand made by it to his reflector. He is, however, now so much out of love with his instrument that he begins to doubt whether it *deserves* a new stand." To a friend, he wrote, "I do not suppose there are many persons who could ever find a star with my power of 6,450, much less keep it, if they found it. Seeing is in some respects an art which must be learnt. To make a person see with such a power is nearly the same as if I were asked to make him play one of Handel's fugues upon the organ. Many a night I have been practising to see, and it would be strange if one did not acquire a certain dexterity by such constant practice."

He returned home from London with a newfound appreciation for his talents as both observer and engineer. One night, writing in his journal, he even allowed himself an uncharacteristic outburst of pride: "I never saw so well, the night is beautiful   my Telescope is the best in the world."

It was true that because reflectors tarnished easily, they required far more maintenance than refractors. Herschel, however, was nothing if not methodical and patient, and he didn't mind the tedium of polishing, especially if his sister read to him *Don Quixote* and the *Arabian Nights,* from Sterne and Fielding. It was also true that reflectors didn't lend themselves to the kinds of measurements that refractors did, but precision concerned Herschel considerably less than power. "The great end in view," he wrote to a friend in 1785, "is to increase what I have called *the power of extending into space.*"

The one insurmountable drawback of the reflector, however, was the position of the person doing the observing. Unlike the refractor, which basically shunted light from one end of a shaft to the other, a reflector bounced the image back up, where an observer, of necessity, would be blocking the light. For this reason, reflectors employed a secondary mirror that redirected the image to an eyepiece either at the side of the tube (in the Newtonian version) or back at the base of the tube (in a design by a Frenchman named Cassegrain). Either way, a secondary mirror absorbed some of the precious light. Herschel, however, dispensed with the need for a secondary mirror by subtly tilting the primary mirror so that the light it deflected back to the opening at the top of the tube would land toward the side, where the observer could stand out of the way. This slight angling in the mirror created some distortion in the image, but in the sizes of mirrors that Herschel used, the distortion by comparison was neglible.

This was the foremost advantage of the reflector over the refractor, one that, for Herschel, easily outweighed all the disadvantages: A mirror could be cast at a far larger size than could a glass blank for a lens. When Herschel inquired about buying mirrors, however, he found them prohibitively expensive, so he tried to commission foundries at Bath and Bristol to cast mirrors for him, but they had no experience in working on the scale that Herschel envisioned. He then took matters into his own hands. Inspired by the factories beginning to crowd the English countryside, Herschel converted his house into a foundry. He devised his own formula for a metal mirror strong enough that it wouldn't bend under its own weight, and he eventually employed gangs of men, wearing numbered work shirts (so he could call directions to them), to execute the mixing and grinding and polishing.

After his London debut Herschel never wanted for commissions. King George III placed him on a royal retainer free of duties other than to show the heavens to members of the royal family whenever they asked. This retainer allowed Herschel to stop teaching music (some of his students in fact preferred to set aside their instruments and spend their lessons discussing the heavens) and to concentrate solely on astronomy, constructing telescopes by day, observing by night. The king himself ordered five 10-foot reflectors. Herschel sold two 7-foot models to the Greenwich observatory, a 25-foot model to the king of Spain, and other reflectors of varying lengths and magnifications to the empress of Russia, the emperor of Austria, and many other crowned heads, as well as a significant number of astronomical luminaries. By 1795 he had made 200 mirrors of 7-foot focal length, 150 of 10-foot, and 80 of 20-foot, including the 2 instruments he used most often over the years, a "small" 20-foot reflector, with a 12-inch aperture, and a "large" 20-foot reflector, with an 18.7-inch aperture.

His greatest achievement in telescope construction was a 40-foot reflector with a 48-inch mirror, far and away the largest telescope to that time. The first casting was a disaster. The mold (made of dried horse dung) sprang a leak, and during the cooling the metal cracked in several places. The second casting was even a greater disaster. The furnace developed a crack, leaking 537.9 pounds of molten metal onto the floor, and Herschel and his workers had to run in every direction to escape both the fiery rush and exploding flagstones. Herschel waited a few years before trying again, and this time he was successful. He mounted the speculum inside a sheet metal tube, the tube inside a wooden framework, and the framework on twenty rollers that ran along a low, circular wall of masonry, enabling him to direct the telescope toward any part of the heavens. It immediately gained an international reputation as the "eighth wonder of the world," not just a tourist attraction but an international destination in its own right.

Herschel, however, wasn't simply an amateur who developed an expertise by happening upon a technology that hadn't reached its potential. He was also a theoretician. Like Galileo, his was the right mind in the right place at the right time. In addition to Uranus (which he always insisted on calling *Georgium Sidus,* or the Georgian Star, after his king and patron), Herschel discovered two satellites of that planet as well as two of Saturn. (Caroline, having abandoned her singing career, made a considerable name for herself as a discoverer of comets, finding eight in ten years.) These were notable achievements, but in the end they were incidental to his larger ambition: "A knowledge of the construction of the heavens has always been the ultimate object of my observations."

In 1783, elaborating upon Halley's discovery that at least three stars were not "fixed" in the firmament, Herschel concluded "that there can hardly remain a doubt of the general motion of all the starry systems, and consequently of the solar one among the rest." In addition, not only was the Sun in motion, Herschel wrote, but he'd determined the direction in which it appeared to be heading: the star Lambda Herculi. Moreover, he realized, all these stars, the Sun included, constitute not an infinite system (or else the sky would be ablaze with light), but one with a definite shape, size, and structure. He even figured out an approximation of that structure, by assuming both that the stars are distributed more or less uniformly and that his telescopes could penetrate to the edges of the stellar regions. It didn't even matter if these assumptions were correct.* They were sufficient to inspire Herschel to collect the data over the course of decades that would enable him to endow the existing catalogs of celestial coordinates with a third dimension—he himself often worked from Flamsteed's *Historia Coelestis Britannica*—and even to construct a reasonable model of what the stars might look like: "a great cluster, the Galaxy, whose shape is roughly that of a convex lens."

Finally, toward the end of his life, Herschel came to see a further implication of his investigations into the third dimension. If light travels around six trillion miles per year, and the *nearest* stars were tens of trillions of miles away, and his instruments were capable of capturing the light from stars thousands of times more distant,

* They weren't.

then a curious conclusion was inescapable. "A telescope with the power of penetrating into space," he wrote, "has also, as it may be called, a power of penetrating into time past."

A clockwork universe, indeed. When Herschel discovered a new planet, he seemed to be offering further affirmation for eighteenth-century optimism, the belief that everything is possible. In fact, when Herschel much later found evidence of double-star systems that seemed to revolve around a common center of gravity, he did extend Newton's law from the solar system to the stars, thereby making it truly universal. But the whole of Herschel's work also wound up suggesting a slightly more subtle alternative to the defining belief of the day: In astronomy *anything* is possible. "I have looked further into space than ever human being did before me," Herschel marveled to the poet Thomas Campbell in 1813, when he was seventy-six. "I have observed stars of which the light, it can be proved, must take two million years to reach the earth."

"*Coelorum perrupit claustra,*" read Herschel's epitaph: "He broke through the barriers of the heavens." More than any other astronomer since Galileo, Herschel left an impression on the world. His achievements were not only the sort that other astronomers would appreciate (though in his case they certainly did) but the kind that captured the popular imagination, that changed the common understanding of the universe. Some forty years after the discovery of Uranus, Keats wrote,

Then felt I like some watcher of the skies,
When a new planet swims into his ken.

But it was a young Alfred Lord Tennyson who more accurately captured the spirit of the age when he urged his brother to overcome his shyness with this remedy: "Fred, think of Herschel's great star-patches, and you will soon get over all that."

In the end Herschel never did succeed in his initial goal of determining how near the nearest stars are. Stellar parallax remained a mystery, perhaps to be solved by some future generation blessed with finer instruments. Nor could Herschel even begin to indicate how far the farthest stars are, to say nothing of the limits of the universe. In 1817, in his final major paper, Herschel summarized his life's work:

> By these observations it appears that the utmost stretch of the space-penetrating power of the 20 feet telescope could not fathom the Profundity of the milky way. From the great diameter of the mirror of the 40 feet telescope we have reason to believe, that a review of the milky way with this instrument would carry the extent of this brilliant arrangement of stars as far into space as its penetrating power can reach, which would be to the 2300dth order of distances, and that it would then probably leave us again in the same uncertainty as the 20 feet telescope.

Nearly half a century, thousands of nights, millions of observations, and the most powerful instruments in history after William Herschel first turned his attention to the stars, the end wasn't even in sight.

Part III

*

BEYOND BELIEF

✴

CHAPTER FIVE

More Light

At the turn of the twentieth century the universe was exactly one galaxy big. Nearly three hundred years earlier Galileo had looked through one of the first telescopes and found evidence supporting Copernicus's contention that the Earth and planets revolve around the Sun, and thus was born the idea of the solar system. In the late 1700s William Herschel had built the largest telescopes ever and counted stars in every direction in an attempt to determine the construction of the heavens, and thus was born the idea of the galaxy. Yet the considerable advances in the technology of the telescope in the century since Herschel had failed to yield evidence of celestial bodies past the stars. Astronomers everywhere had started to wonder: Was this it, then, journey's end? Had they truly glimpsed the farthermost reaches of existence? And if that were so—if everything visible through the most powerful instruments lay within the Milky Way galaxy, and the Milky Way galaxy defined the limits of the universe—then what lay *beyond?*

For once, some observers were willing to call it quits. "We have reached a point," wrote the historian Robert Ball in his 1886 work *The Story of the Heavens,* regarding the realm beyond the galaxy, "where man's intellect begins to fail to yield him any more light, and where his imagination has succumbed in the endeavor to realize even the knowledge he has gained." In 1905 Agnes Clerke, the leading historian of nineteenth-century astronomy, called the one-galaxy universe a "practical certainty," and added, "With the infinite possibilities beyond, science has no concern." It was as if the medieval "No farther" that the seers and sages of the late Renaissance had triumphantly transformed into "Farther yet" had passed, at last, into "Far enough." "Of infinity in any of its aspects we can really know nothing," Alfred Russel Wallace, who several decades earlier had arrived at the theory of evolution at the same time as Charles Darwin, concluded in 1903. "To me its existence is absolute but unthinkable—that way madness lies."

George Ellery Hale talked to an elf. According to one enduring legend, the elf first visited him during a recuperative stay in Egypt in 1910, following one of Hale's periodic "nervous breaks." The elf returned a few weeks later, in Rome, hectoring Hale to put down the book he was reading and get back to work. "How to escape this new form of torture, which is incessant, I do not know," Hale wrote to a friend on that occasion, but over the years, as the elf continued to visit him, and as he withdrew from public life, Hale came to regard these intrusions as a peculiar kind of companionship. There is no evidence that the elf was a result of Hale's frequent excursions into the infinite, just as there's no way to prove that Galileo went blind from staring into the Sun. But for the astrophysicist and visionary who oversaw the creation of four telescopes that ranked among the largest in the world—who was (apparently painfully) aware that he was the last, best hope for astronomers who still wanted to know just where the end of the universe is—an elf with answers might do the trick.

Hale's feet tingled, his head ached, his ears rang, and his nightmares sent him literally up the walls. ("Sometimes he would get up in the night and in his tormented half sleep would try to climb the picture frames," one friend recalled.) He enjoyed periods of clarity, of productivity, of tremendous energy and accomplishment, and he suffered episodes of debilitating depression and isolation. He was prone, he confessed, to "over-mobilization," which taxed his "busted old head." But he couldn't help himself. When seized with a vision for a new machine that could reach farther into space and farther across time than his previous one, he wouldn't stop working on it until he had willed it into existence, at which point he would find himself supervising every detail of its daily operations while simultaneously laying out elaborate plans for his next new, best machine. "More light!" he would cry to the members of his staff, and then he'd go out and get it.

His first major telescope was a 40-inch-diameter refractor, which opened in 1897 at the Yerkes Observatory at the University of Chicago. It had a light-gathering power of 35,000 times that of the human eye.

A decade later Hale mounted a 60-inch mirror at Mount Wilson, near Los Angeles, with a light-gathering power 57,600 times that of the human eye. A decade after that Hale delivered another mirror to Mount Wilson, this time with a diameter of 100 inches and a light-gathering power of 160,000 times—sufficient as to allow a human eye to see stars as faint as the nineteenth magnitude, or a candle at a distance of 2,400 miles, and to allow a photographic plate to detect stars of the twenty-first magnitude, or a candle at 9,600 miles. And still, yet another decade later, Hale continued to call out, more quietly now because of the ravages of his mental illness, though with no loss of conviction, "More light!"

Over the course of the nineteenth century, astronomers had learned the lessons of Herschel. It was some measure of just how thoroughly they'd done so that by Hale's era a telescope was known not by its focal length but by the diameter of its aperture, not by how much it magnified an image but by how well it gathered light. The Yerkes telescope, for instance, was never a 63-foot, always a 40-inch, refractor. Magnification remained crucially important, but in many ways the story of telescopic development in the century from Herschel through Hale was one of a shift in emphasis, of the ascendance of a new priority: of a growing appreciation for the meaning of more light. More light meant more depth, and what more depth meant, astronomers quickly learned, was more nebulae.

A nebula—from the Latin for "cloud"—was one of the faint, unidentifiable smudges in the night sky, neither planet nor star, or either, or both. Before the invention of the telescope, astronomers had noted the existence of nine such objects. Between Galileo and Herschel, the number of nebulae had risen to ninety. By the time Herschel was done with them, they numbered twenty-five hundred. In December 1781 Herschel had received a catalog of sixty-eight nebulae, published the preceding year by one Charles Messier, an astronomer who specialized in discovering comets and who had compiled a list of nebulous objects in the heavens because they kept getting in the way of his true calling. Herschel, however, found nebulae fascinating objects of study in their own right. When he wasn't counting stars, he was often trying to classify nebulae. In the end he never was able to determine whether nebulae were single stars surrounded by a shell of some sort, compact clusters of stars, a mysterious gaseous matter, or even, in the terminology of the times, "island universes"— vast collections of stars wholly separate from, and equal in size to, our own Milky Way galaxy. All he knew for sure was that the farther he looked, the more he found, often in new and puzzling shapes.

The wide reflecting telescope and the search for nebulae had reached a new extreme in the decades after Herschel's death. In 1845 an Irish nobleman, William Parsons, the third earl of Rosse, completed a reflecting telescope that became known simply as the Leviathan of Parsonstown: a 72-inch mirror weighing four tons, at the base of an open 54-foot tube, which hung from chains between two walls of stonework more than 50 feet in height. The observer would perch on a precarious platform, high above the ground, and peer down the tube. Not surprisingly, the telescope wasn't especially mobile. Its location—the earl's castle grounds on the wet, windswept island of Ireland—suffered from the kind of local atmospheric conditions that astronomers call, with admirable bluntness, "bad seeing." In addition, Rosse's charitable work during the potato famine led him to suspend operations for several years after the first observations. Despite these setbacks, the Leviathan did make one significant contribution to astronomy. In the spring of 1845 Rosse examined the nebula 51 Messier (from the numbering system in the catalog that had helped inspire Herschel) and found that it had a curious shape, a spiral. Over the coming decades Lord Rosse found more nebulae of the same shape, to such an extent that astronomers began to wonder whether most nebulae would prove to be spirals, if only their own telescopes were sufficiently powerful to grasp enough light.

For the reflector had reached the limits of its technology, for reasons that William Herschel himself had anticipated. In many ways, Herschel's great 40-foot telescope with its 48-inch mirror, for all its renown, had disappointed its creator. It too was cumbersome to manipulate. Two men were required to maneuver it, and even then a break in the weather meant a race against time; a 20-foot telescope, by contrast, took one man ten minutes to ready. Because of the extraordinary difficulty in fashioning a speculum 4 feet in diameter, the telescope had no mirror in reserve; when the mirror needed polishing, the telescope simply went out of commission—and the mirror needed a great deal of polishing. In order to keep the mirror from cracking under its own weight, Herschel had toughened the alloy by adding copper, which caused it to tarnish far more quickly than the relatively more lightweight mirror in a 20-foot telescope. "A 40-foot telescope," Herschel once wrote, "should only be used for examining objects that other instruments will not reach. To look through one larger than required is loss of time, which, in a fine night, an astronomer has not to spare." Over the decades he used the 40-foot less and less and polished it more and more. "The mirror is extremely tarnished," he noted one night in 1815, and never used it again.

At this point the history of the telescope paused for one of those periodic impasses where the technology can't satisfy a demand for new information. This time, however, the breakthrough came relatively swiftly. In 1856 a German mathematician and a French physicist separately arrived at a way to deposit a thin, even layer of silver on glass and achieve the same reflecting results as speculum metal. This new reflecting process produced mirrors that were not only lighter than those made with speculum metal, and therefore easier to mount and maneuver, but more effective. Silver-on-glass mirrors, it turned out, reflected half again as much light as metal.

In terms of precision work, however, it was the *refractor* that dominated professional astronomy throughout much of the nineteenth century. A triple-lens eyepiece combination, developed in the mid-1700s, had finally eliminated chromatic aberration (as well as the need for absurdly long, unwieldy refractors), while other improvements in grinding and glass purification techniques had significantly diminished spherical aberration. In fact, by the late 1830s the quality and size of glass lenses had improved to the extent that three astronomers separately arrived at the long-elusive goal of determining the distance to a star. The honor of priority went to Friedrich Wilhelm Bessel, who detected a parallactic shift in 61 Cygni, a star at the limit of naked-eye visibility, that he calculated would place the star at a distance of 657,700 astronomical units, or approximately 61 trillion miles. Astronomers now had an accurate scale of distance from the solar system to the stars, as well as a new testimonial to the refractor's capacity for precision: The difference in position that Bessel had detected over a six-month period was a third of an arc second, or the equivalent of 2 feet at 260 miles.

But in terms of size—and therefore the capacity for gathering light—refractors could compete with the new silver-on-glass reflectors only up to a point. Even the largest, most unwieldy telescopic mirror can rest securely at the bottom of the instrument, but a glass lens has to hang at the top of the tube, bearing much of its own weight. If it sags, it's useless. A thicker lens was a stronger lens, of course, but it also was a more light-absorbing (rather than light-refracting) lens, and beyond a diameter of about a meter a glass lens would need to be so thick that it would absorb more light than a silver-on-glass mirror of the same size. Over the course of the nineteenth century, astronomers came to understand that the superior light grasp of the reflector offered them the equivalent of a new technology, and it eventually became the telescope of choice at most major astronomical observatories if only so astronomers could see what it could do.

In a way the quest for more light had returned modern astronomy to its telescopic roots. In its long refractor incarnation the telescope had become a quantitative instrument perfect for completing the calculations of a cause-and-effect Newtonian cosmos. The quest for more light that Herschel had initiated, however, wasn't necessarily a search for numbers. Like the first astonished readers of *Sidereus Nuncius* clamoring for a peek through a *perspicillum* to see what Galileo had seen, astronomers of the nineteenth century wanted to see what Herschel had seen, and once they did, they wanted to see *more.*

More light meant more depth, and more depth meant more nebulae. But what did more nebulae mean? Specifically, the questions that the search for more light raised were twofold: What are the nebulae? Where are the nebulae?

The search for answers to those questions began, appropriately, in William Herschel's backyard. In 1839 John Herschel, like his late father an astronomer and composer, bade farewell to the famous, abandoned 40-foot reflector. He removed the mirror, dismantled the decrepit wooden mounting, and laid the tube to rest on the lawn. Then he gathered his wife, children, and a small band of musicians inside the cavernous space and led them in singing "Requiem of the forty feet reflector at Slough: Sung on New Year's Eve, 1839–1840":

> *In the old Telescope's tube we sit,*
> *And the shades of the past around us flit,*
> *His requiem sing we with shout and din,*

While the old year goes out and the new
   comes in.
Merrily, merrily, let us all sing,
And make the old telescope rattle and ring.

The telescope was down but not, as it turned out, gone. A few months earlier, before he'd begun dismantling the telescope, John had preserved the memory of the one-time wonder of the world by committing its image to the first photograph on glass.

John Herschel had been experimenting with physical optics and chemistry for twenty years, and he was one of the genuine pioneers in photography. "Photography," in fact, was among his contributions to the vocabulary of the new technology, along with "positive," "nega-tive," and "snapshot." As with the optic tube in its early days, pointing a light-sensitive device at a dark sky might not have been the first use to occur to novices, but it soon proved to be an irresistible one. In 1840 the American chemist Henry Draper succeeded in captur-ing the first, albeit faint, daguerreotype—a highly pol-ished, iodine-fumed silver plate—of the Moon, through a twenty-minute exposure. Over the coming decade, other celestial photographic firsts followed: a solar eclipse, the Sun itself, stars. The director of the Paris Observatory, an early dabbler in daguerreotypes, fore-saw a time when photography would help even in map-ping the moon and determining the magnitudes of stars. "After all," he wrote, "when observers apply a new instrument to the study of nature, what they had hoped for is always but little compared with the succession of discoveries of which the instrument becomes the source—in such matters, it is on the unexpected that one can especially count."

In its earliest incarnations what photography offered astronomers primarily was convenience, especially once technical improvements had helped to render glass plates obsolete and to increase light sensitivity, reducing exposure times by a factor of twenty or thirty. "On a fine night," an astronomer at Harvard University wrote to a friend, "the amount of work which can be accomplished with entire exemption from the trouble, vexation and fatigue that seldom fail to attend upon ordinary observations, is astonishing. The plates, once secured, can be laid by for future study by daylight and at leisure. The record is there, with no room for doubt or mistake as to its fidelity." Productivity, precision, certainty: What more could an astronomer want?

Light, as always. Just as the telescope itself turned out to be an instrument capable of more than magnification, so the camera's significance for astronomy ranged beyond simply capturing an image. Like the telescope, it gathered light—more light than the human retina can absorb. In 1882 David Gill, her majesty's astronomer at the Royal Observatory, Cape of Good Hope, borrowed a camera to take a picture of a comet and found in the resulting photograph a host of stars in the background that he hadn't noticed at the time—stars that, as far as his eyes were concerned, hadn't even been there at the time. Afterward he devoted himself for five years to taking pictures of the southern sky, dividing the hemisphere into a total of 612 fields, each requiring an exposure time of half an hour to an hour. The resulting catalog, the *Cape Photographic Durchmusterung,* contained all the southern stars up to the visual magnitude of 9.5—a total of 454,875 stars, and a convincing testimonial to telescopic photography's ability to produce an enduring and exhaustive record.

But unlike the telescope, a photograph didn't *just* gather light. It didn't simply collect the light that happened to be present at a particular moment. Rather, it gathered light, and it *kept* gathering light. Light poured onto the film stock and stayed there—and the longer the exposure, the more light that accumulated. In 1888 a three-hour exposure by the British astronomer Isaac Roberts, using a 20-inch reflector, revealed for the first time the spiral nature of the great nebula in Andromeda. Later that same decade the French brothers Paul and Prosper Henry pointed a photographic telescope at the Pleiades, opened the shutter for hours, and later counted—in the same region where Galileo in *Sidereus Nuncius* had triumphantly added 40 new stars "to the six or seven we can see"—2,326 stars. And that wasn't even their most striking result. Here, in a region of sky that astronomers had studied since the beginning of astronomy, they also found a new nebula.

In itself the capacity for accumulating light would have made photography indispensable for telescopic astronomy. In 1900 James Keeler, the director of the Lick Observatory in California, stunned his fellow astronomers when he estimated the number of nebulae within reach of his 36-inch reflector as 120,000, and he added that they "exhibit all gradations of apparent size, from the great nebula in Andromeda down to an object which is hardly distinguishable from a faint star disk." This work in particular helped validate the use of cameras by professional astronomers. As much light as a telescope of great aperture could gather, clearly a telescope and a camera could accumulate much more.

But a telescope, a camera, *and* an instrument called a spectroscope could change the very nature of astronomy. In the 1830s the French positivist philosopher Auguste Comte had argued that what is unknowable is not worth pursuing, and he cited as an example the composition of celestial bodies—that mysterious fifth element that Aristotle had called quintessence, and that modern astronomers, for all their progress over the centuries, were no closer to identifying. "We conceive the possibility of determining their forms, their distances, their magnitudes, and their movements," Comte wrote, "but we can never by any means investigate their chemical composition or mineralogical structure." Apparently taken with his argument, he repeated it years later, only with greater emphasis: "It is then in vain that for half a century it has been endeavored to distinguish two astronomies, the one solar, the other sidereal. In the eyes of those for whom science consists of real laws and not of incoherent facts, the second exists only in name, and the first alone constitutes a true astronomy; and I am not afraid to assert that it will always be so."

Comte died in 1857. Two years later two physicists used a spectroscope to determine the chemical composition of the Sun. Five years after that, an astronomer used a spectroscope to figure out what a nebula was.

The spectroscope had been invented in 1815 by the great German spectacle maker Joseph Fraunhofer when he duplicated Newton's 1666 experiment of passing light through a prism, but with a difference: First Fraunhofer directed the beam of sunlight through a telescope. The resulting magnified spectrum turned out to be not the one continuous, imperceptible blend of colors, violet through red, that Newton had found, but "an almost countless number of strong and weak vertical lines." Fraunhofer counted these lines, charted them, and in the end identified more than five hundred of them in the spectrum from the Sun, though he never could identify just what those lines signified.

What they signified, the German chemist Robert Wilhelm Bunsen and physicist Gustav Robert Kirchhoff figured out in the 1850s, was the chemical constitution of the matter producing them. Different chemicals had different patterns of lines, and each chemical had its own distinctive pattern. Since chemists could match patterns with chemicals, the composition of celestial bodies was there for the taking, at least in theory.

In 1863 the British astronomer William Huggins turned this instrument toward a nebula. According to Bunsen and Kirchhoff, if Huggins were to see a continuous, multihued, line-riddled spectrum, the nebula in his sights would contain a star or stars, an unbroken straight line, glowing gas. Huggins later recalled the moment of discovery:

> On the evening of August 29, 1864, I directed
> the telescope for the first time to a planetary

nebula in Draco. The reader may now be able
to picture to himself to some extent the feeling of excited surprise, mingled with a degree
of awe, with which, after a few moments of
hesitation, I put my eye to the spectroscope.
Was I not about to look into a secret place of
creation?

I looked into the spectroscope. No spectrum
such as I expected! A single bright line only!

Huggins double-checked the spectroscope. The
equipment was sound, the evidence apparently real.
"The riddle of the nebulae was solved. The answer,
which had come to us in the light itself, read: Not an
aggregation of stars, but a luminous gas."

In fact, the riddle of the nebulae *wasn't* solved—only
the riddle of this one nebula. In the next two years
Huggins examined a total of sixty nebula and found that
a third showed the bright-line spectrum that denotes a
glowing gas, or, in William Herschel's term, "true nebulosity," while the rest showed the continuous spectra
that signal the presence of a star or stars. But the experiment with the nebula in Draco did mark the birth of a
new branch of science, astrophysics, and a new kind of
astronomy—indeed, its adherents argued, a New
Astronomy.

Individually, either photography or spectroscopy
could have proven to be a powerful new astronomical
tool, capable of yielding information that the telescope
alone could not. Together they represented a scientific
discipline greater than the sum of its parts. If you pass

the faint light from a star, or the even fainter light from a nebula, down a telescope and through a prism, the resulting spectral lines could easily be illegible. But if you pass the same light down a telescope, through a prism, *and* onto a photographic plate, you could let it accumulate there until the resulting spectral lines would be distinct enough to read. In 1872 Henry Draper used a 28-inch reflector on Vega to determine the first photographic spectrum of a star, and from that day forward an astronomer could reach seemingly anywhere in the universe and walk away with a record of what was there— not just what it looked like, but what it was.

And what it was was us. The continuous spectra from certain celestial bodies—the Sun, other stars, some nebula—turned out to correspond to the spectra from familiar terrestrial chemicals. "The characteristic light rays from earthly hydrogen shone side by side with the corresponding radiations from starry hydrogen," Huggins wrote. "Iron from our mines was line matched, light for dark with stellar iron from opposite parts of the celestial sphere. Sodium which upon the earth is always present with us was found to be widely diffused through the celestial spaces."

Galileo had used the telescope to penetrate the heavens and see that the heavenly bodies and Earth share the same physical appearances. Newton had theorized that celestial and terrestrial objects share the same laws of physics. Now Huggins had shown that they *are* the same.

"Then it was that an astronomical observatory began, for the first time, to take on the appearance of a laboratory," Huggins wrote.

Primary batteries, giving forth noxious gases, were arranged outside one of the windows; a large induction coil stood mounted on a stand on wheels so as to follow the positions of the eye-end of the telescope, together with a battery of several Leyden jars; shelves with [Robert Wilhelm] Bunsen burners, vacuum tubes, and bottles of chemicals, especially of specimens of pure metals, lined its walls.

The observatory became a meeting place where terrestrial chemistry was brought into direct touch with celestial chemistry.

The New Astronomy, as the deliberate evocation of Kepler's formative 1609 work suggests, marked a break with the past. The New Astronomy of the seventeenth century had departed from tradition by focusing not on mathematical attempts to save the appearances but on the appearances themselves: how the celestial bodies look and move. Now the New Astronomy of the nineteenth century was shifting the emphasis yet further: what the highly influential 1888 best-seller *The New Astronomy* defined as the "Sun, Moon, and stars for what they are in themselves, and in relation to ourselves." First the telescope had brought the heavens tantalizingly close—seemingly close enough to touch. Now, through the telescope, they actually were; we actually could.

In a way, George Ellery Hale's lifelong search for more light was a practical matter. More light was what the New Astronomy demanded—photography and spectroscopy, alone or in tandem. Not only did telescopes with larger apertures penetrate more deeply into the region of the nebulae, but they dispersed the closely packed lines in spectra more widely and therefore more legibly. "You know it is not altogether absurd to regard a telescope as simply a large lens used to form an image of a celestial source of light on the slit of a spectroscope," Hale wrote while planning his first major telescope, for the University of Chicago. "In this sense the whole institution may be regarded as a physical laboratory." Hale oversaw the construction of a 40-inch that was then the largest refractor in the world,* as well as a sprawling, multiacre institution unlike any other at the time. When it opened in 1897, in Williams Bay, Wisconsin, the Yerkes Observatory epitomized the New Astronomy—laboratories for optical, spectroscopic, and chemical work; darkrooms; developing, emulsion, and enlarging rooms—and Hale later heard from no less an eminence than Sir William Huggins: "It must indeed be of untold sati[s]faction to you to be at the head of so magnificent an Institution for combined astronomical and laboratory work! It seems to me almost marvelous that within a single lifetime, my first taking a simple spectroscope into an Observatory has borne such magnificent fruit in both hemispheres."

Hale's constant quest for more light, however, was no less a philosophical matter—a mission, even. In 1889 at the age of twenty-one, Hale invented the spectrohelio-graph, a device for photographing the Sun. In the early 1890s he helped found the journal *Astronomy and*

* It still is.

Astrophysics, the precursor to *The Astrophysical Journal: An International Review of Spectroscopy and Astronomical Physics,* and received a letter from the editor of the influential British journal of science *Nature:* "It is by means of such a journal as yours that the cause of the New Astronomy will be advanced." In 1899 he was instrumental in founding the American Astronomical and Astrophysical Society; the "Astrophysical" part of the organization's title was added at his insistence, and he allowed it to be dropped, fifteen years later, only when he believed that the New Astronomy's status was secure.

As in the battle between Johannes Hevelius and John Flamsteed over the use of telescopic sights, astronomers first had to decide whether they could trust the new information. Many didn't—like Hevelius, often with good reason. The technology was still raw; the results were often unreliable. Just as an earlier transitional generation of astronomers needed to specify whether an observation had been naked-eye or telescopic, so the new New Astronomers needed to designate whether an observation was "visual" or "photographic." Between 1881 and 1887 an international congress of science practitioners convening in Paris recommended forgoing the use of photography in astronomical research. As late as 1914 the American astronomer Percival Lowell could characterize astrophysics as that science "which, of late years, owing to the effect that pictures have on people, has usurped to itself the lime-light to the exclusion of the deeper and more profound parts of astronomy proper."

Yet it was the form this new information took—"the effect that pictures have on people"—that distinguished it from all earlier research. For the first time astronomical evidence resided not in the eye of the beholder, and not in the instrument, but at one further remove: the output from the instrument. Astrophysics produced evidence that theoretically was the same for everyone because it could exist independently of the observer who gathered it—could even outlive the original observer. An astronomer at Princeton Observatory in 1900 praised the New Astronomy for providing precisely this: "a record that is permanent, authentic and free from the personal bias of imagination and hypothesis, which so seriously impairs the authority of many ocular observations."

In a way this removal of the observer from the equation—the attainment of objectivity—might have seemed a fulfillment of the philosophical promise of the previous century: the universe of Isaac Newton as interpreted by Pierre Simon Laplace, one where the answers were out there waiting to be discovered, and it was only a matter of when. In fact, by the end of the century two pieces of evidence had emerged to make a strong case that, having determined what the nebulae were, astrophysicists were ready to determine where they were—specifically, whether the nebulae were part of our Milky Way galaxy or were separate island universes of their own.

First, a nova, or new star, appeared in the Andromeda nebula in 1885 that was one-tenth as bright as the entire nebula. If Andromeda were indeed an island universe equal in size to the Milky Way galaxy, then the nova

would have to have been as bright, according to the estimate of one leading historian of the time, as "nearly *fifty
million* such suns as our own"—a scale "outraging all
probability."

Second, as William Herschel had noted a century earlier, and as recent observations confirmed, the nebulae
seemed to cluster away from the whitish stretch of sky
commonly called the Milky Way, an area that nebular
investigators now began calling the "zone of avoidance." If the nebulae were indeed numerous other galaxies, they couldn't have arranged themselves as if to
avoid a segment of the sky in our galaxy. Then again, if
they *were* within the galaxy, it would make sense that
they would be arrayed away from the center, near the
edges—and therefore, from our point of view, at the farthest reaches of the universe and the farthest back in
time, in some earlier stage of stellar development.

In 1858 the English philosopher Herbert Spencer had
written of the existence of nebulae beyond our galaxy,
"Such a belief is next to impossible." By 1899 he was
ready to revise this view: "Such a belief is impossible."
The leading American astronomer at the time, Simon
Newcomb, wrote in 1902 that "the great mass of stars is
included within a limited space." And Alfred Russel
Wallace asserted in 1903 that this "grand and far-
reaching principle of the essential unity of the stellar
universe . . . is now accepted by almost every astronomical writer of eminence in the civilized world."

"The question whether nebulae are external galaxies
hardly any longer needs discussion," the historian
Agnes Clerke wrote in 1905. "It has been answered by

the progress of research. No competent thinker, with the whole of the available evidence before him, can now, it is safe to say, maintain any single nebula to be a star system of co-ordinate rank with the Milky Way."

Having settled what and where the nebulae were, most astronomers figured that the only remaining tasks were tying up such loose ends as assigning positions to the hundreds of thousands of new nebulae that were streaming onto the photographic plates of the most powerful telescopes and one day, technology willing, even determining their distances. "While it is never safe to affirm that the future of Physical Science has no marvels in store even more astonishing than those of the past," read a passage in the University of Chicago catalog for the 1898–99 academic year, "it seems probable that most of the grand underlying principles have been firmly established and that further advances are to be sought chiefly in the rigorous application of these principles to all the phenomena which come under our notice. An eminent physicist has remarked that the future truths of Physical Science are to be looked for in the 6th place of decimals."

At the same time and at the same institution, Hale was opening the Yerkes Observatory—and what was Hale's own constant striving for something bigger, better, deeper, farther, further, newer if not a belief in progress? He even had a name for his condition: "Americanitis." The machinery of a modern observatory could match the mechanics of a clockwork Creator cog for cog, gear for gear. William Herschel's gang of number-jerseyed workmen had grown into the factories and assembly lines of the Industrial Age, and when Hale wanted to evoke comparisons for his observatories among construction and engineering feats, he turned to battleships and bridges. Major observatory telescopes now demanded domes to protect their mechanisms from the elements and needed the stability of piers sunk into bedrock independent of the observatory floor. At Yerkes, the 40-inch refractor stretched 63 feet into the air and weighed 6 tons, with the gears to move it adding another 20 tons; the dome spanned 90 feet in diameter and weighed 120 tons; and the 37-ton movable observatory floor operated independently of the telescope, raising the observer to the eyepiece. Yet the entire enterprise required nothing so much as a great delicacy, the grace to follow the gliding movement of a celestial object across the sky for hours at a time without blurring a photographic exposure, the subtlety to respond to a fingertip touch.

This level of technology didn't come cheap, but here, again, Hale was a man of his times. His father had made a fortune after the 1871 fire in Chicago by supplying elevators for the new skyscrapers, and Hale never hesitated in approaching the barons of industry for enormous sums of money, often promising little more in return than gilt by association. In this, he was following a precedent established at the Lick Observatory, the beneficiary of a bequest by an eccentric California millionaire who wanted his name on a telescope "superior to and more powerful" than any other; in the end old man Lick's remains were interred in the pier supporting the telescope. For his University of Chicago observatory, Hale turned to a mass transportation mogul and convicted swindler. YERKES BREAKS INTO SOCIETY, read one Chicago headline after the dedication of the observatory. "Street Car Boss Uses a Telescope as a Key to the Temple Door and It Fits Perfectly." Hale named the 60-inch Hooker reflector after a Los Angeles hardware and steel pipe magnate, and Andrew Carnegie once wrote to Hale regarding his own contribution on behalf of the 100-inch reflector, "I hope the work at Mt. Wilson will be vigorously pusht, because I am so anxious to hear the expected results from it. I should like to be satisfied before I depart that we are going to repay to the old land some part of the det we owe them by revealing more clearly than ever to them the new heavens."

"I wished Carnegie could keep his millions to himself," Hale's wife once observed when his worries about the 100-inch mirror interrupted their vacation by pitching him into an anxious frenzy and, for the first time, manifested themselves as an elf. "I wish that glass was in the bottom of the ocean." Her husband was sympathetic but unrepentant. It was one thing to believe that the edge of the universe might be in sight and another to be the only person capable of ensuring that anybody would ever get a chance to see it. Once, when his wife suggested that he attend church on Sunday for the sake of the children, Hale answered, "My creed is truth, wherever it may lead, and I believe that no creed is finer than this."

It was this creed that distinguished him from many of his contemporaries. For Hale more light meant more

answers, but not necessarily of the "6th place of decimals" variety. It was true that Hale fully embodied the
optimism, the faith in progress, that he'd inherited as a
New Astronomer at the turn of the twentieth century,
but it was also true that he was unwilling to reach the
limits of the galaxy and call it quits. He hadn't glimpsed
the end, either of the universe or of progress; why
should he believe he'd reached it? Before astrophysicists could decide what they wanted to do with their
new technology, Hale wanted to see where *it* led *them*.

Still, not even he, for all his New Astronomy prose-
lytizing and Industrial Age philanthropizing, could have
remotely anticipated the truths that would be passing
through the portals of the modern observatory or how
the astronomers watching there would be interpreting
them. In 1917 Heber D. Curtis, of the Lick Observatory,
rebutted the two supposedly decisive arguments against
the existence of "island universes." First, in examining
photographs of spiral nebulae, he noticed that many of
them exhibited a "dark lane down center," a thin band
of obscuring matter, some sort of cosmic dust. If the
Milky Way were a spiral galaxy, then couldn't a similar
dark lane account for the "zone of avoidance," rather
than an actual absence of nebulae? Second, Curtis also
noticed evidence of novae on photographs of spiral nebulae, but none approaching the magnitude of the 1885
nova in Andromeda. Maybe it *had* been abnormal—
meaning that the universe just might be capable of dis-
plays matching the output of fifty million Suns after all.
The argument against the presence of extragalactic nebulae was starting to sound like the one that anti-

Copernicans had mounted regarding the absence of a measurable parallactic shift in the stars—that it rendered the distance to the stellar sphere unimaginable, if not impossible—with much the same result. It *was* unimaginable; it *wasn't* impossible.

At the same time, other evidence was beginning to mount that didn't just counteract the one-galaxy argument but actively supported the island universe hypothesis. In 1868 William Huggins had speculated that the lines produced by a celestial object in a spectroscope would exhibit an effect similar to the one that Christian Doppler had noticed in sound waves: a shift, depending on whether the object making the lines or waves is moving away from or toward the viewer or listener. Specifically, Huggins suggested, a nebula exhibiting a shift in spectral lines toward the blue part of the spectrum will be moving toward the observer; a nebula exhibiting a redshift will be moving away—and the greater the shift, the greater the velocity.

Shortly after the turn of the century Vesto Slipher, of the Lowell Observatory in Arizona, discovered several such shifts. Moreover, the shifts he recorded indicated that pieces of the universe were moving at a velocity that hardly seemed credible for anything, let alone massive congregations of millions of stars. In 1913 he calculated that the Andromeda nebula was rushing through space at three hundred kilometers per *second*.

By 1920 Hale had taken it upon himself to organize a "debate" at the National Academy of Sciences on the issue of where the nebulae were. Most in attendance agreed that the Lick Observatory's Heber Curtis, arguing on behalf of the multiple galaxy universe, "won" on rhetorical grounds, but as Hale himself later wrote, "It is a far cry from the facile imaginings of the philosopher to the rigorous demonstrations of exact science, and the true structure of the universe is not yet known." Argument alone wasn't going to settle what was becoming the defining astronomical issue of the young century; only evidence was.

Three years later Edwin Hubble, a colleague of Hale's from Mount Wilson, found it. Since his arrival at Mount Wilson in 1919, he had been using the 100-inch to search the nebulae for a class of star called a Cepheid variable, a star with regularly varying luminosity. According to recent findings, the period of light variation in a Cepheid variable would correspond to the absolute brightness of the object. By comparing the absolute brightness (how bright the Cepheid variable really is) with the apparent brightness (how bright it appears to an observer on Earth), an astronomer could apply a simple mathematical formula to produce the distance of the object. Hubble figured that by identifying a Cepheid on a photographic plate of a nebula, he would be able to determine the distance of the entire nebula. On October 6, 1923, his search ended, in the Andromeda nebula. He spent another year confirming his finding, eventually estimating that Andromeda lay about a million light-years, or six quintillion miles, away—far outside the farthest reaches of the Milky Way.

So our galaxy wasn't the universe in its entirety after all. Some of the nebulae that astronomers had discovered over the preceding century displayed true nebulosity, and some were star clusters, and some were single stars surrounded by a shell of some sort—and all of these presumably lay within our galaxy. What lay beyond were all the *other* nebulae—those hundreds of thousands of spirals on the Lick Observatory's photographic plates, for instance, and each very likely comprised of hundreds of millions of stars.

As recently as the turn of the century most astronomers believed not only that the universe consisted of exactly one galaxy, but that within this galaxy our Sun occupied a position that, though appreciably less distinctive than the one it had held when the stellar region still resembled a sphere or vault, was central. Now astronomers had determined not only that the local star seemed to be lying somewhere significantly more toward the perimeter of the galaxy than the center, but that the galaxy itself was only one of an unknown but doubtless staggering number of such massive star systems.

Hubble, however, wasn't done. In 1929 he announced that in determining the distances to twenty-four nebulae outside our galaxy—these galaxies in their own right—he had detected a pattern. It was a simple relationship with a potentially profound implication: The farther the galaxy, the greater its redshift, or velocity. Although Hubble was careful not to endorse the obvious conclusion himself, the implication was unmistakable. In every direction, at all times, the universe was expanding.

Already Hale, one eye on history, one eye on posterity, was busy working on a mirror that would reach yet farther across the universe, look yet deeper into the past. "Lick, Yerkes, Hooker, and Carnegie have passed on,"

Hale had written in *Harper's Magazine* in 1928, hinting for help in underwriting his next new, best reflector, "but the opportunity remains for some other donor to advance knowledge and to satisfy his own curiosity regarding the nature of the universe and the problems of its unexplored depths." Shortly after this plea appeared, Hale received a summons from the Rockefeller Foundation. "An article of mine on large telescopes, shot like an arrow into the blue, seems to have hit a 200″ reflector," Hale wrote to an associate soon afterward.

By now Hale had resigned his directorship at Mount Wilson owing to his increasingly fragile health. For weeks at a time he would disappear into his private solar study. Sometimes he would take the long train trip across the country and retreat to a sanitarium in Massachusetts. During one such stay he wrote to his wife that she was right to object to his "high tension and great interest" in multiple projects and that he hoped to make amends. "This I must try to overcome," he went on, "but as it was born in me and has been exercised so long I may have to work for a long time before I succeed."

Yet already he'd agreed to oversee the design, organization, and construction of what was sure to be his final major project, a 200-inch telescope atop California's Mount Palomar. "The last thing I had in view," he wrote a fellow astronomer on his return from the sanitarium, "was to create a new observatory." But how could he resist? A "whirligus" of his own design and beyond his control was once again fully upon him. What else was a man like George Ellery Hale to do—let the universe go on expanding without him?

*

CHAPTER SIX

More Dark

Hiss.

Karl Jansky heard about it before he heard it. It was a common sound for an engineer to encounter, a sound like air leaking from a tire, and on this occasion it was a sound that happened to be interfering with the newly opened transatlantic radiotelephone transmissions. Jansky, twenty-six, had been assigned to investigate the source of the hiss by the Bell Telephone Laboratories in Holmdel, New Jersey, and he constructed a set of separate aerial antennas, thirteen feet tall, thirteen feet deep, and, taken together, one hundred feet long. He mounted the entire array on four wheels he'd appropriated from a Model T Ford, called it a merry-go-round, and set it rotating. As the antennas turned toward different sections of the sky, he tried to pinpoint the various origins of the background static. He identified the familiar crackle of nearby storms and the equally familiar perpetual crackle of the Earth's atmosphere gathering itself into distant storms, but this other, persistent hiss remained a mystery. He thought it might be coming from local electrical equipment, but further study showed that the source seemed to be progressing in daily increments, slowly but steadily, along the horizon. So he thought it might be coming from the Sun, but a partial eclipse on August 31, 1932, didn't diminish it. This, then, was no ordinary hiss. This was a hiss without cause and without cure. It was everywhere, and it was always there. It was the hiss of the snake in the Garden, the whisper in the ear of Pandora, the air going out of the universe.

It was, Jansky wrote, more prosaically, a "steady, weak, hiss-type static of unknown origin." The key to the mystery seemed to be its movement. If it wasn't coming from the Sun, then why did the position of the source of the hiss seem to change slightly every day, as if it were following the solar calendar? Jansky refined his measurements and found that the cycle of the hiss actually wasn't twenty-four hours, but four minutes less. That is, its cycle was twenty-three hours, fifty-six minutes, precisely the length of the sidereal, rather than the solar, day—the time it takes the stars to make one revolution in the sky.

That was it, then. The reason the hiss seemed to be coming from everywhere at once, and all the time, was that it *was*. The source was the stars themselves, Jansky figured, and in time he pinpointed the greatest concentration of static as coming from Sagittarius, or the direction of the center of the Milky Way galaxy. He wanted to continue his research, but he had accomplished his mission, and Bell Labs reassigned him. Jansky published his results in *Popular Astronomy* and *Proceedings of the Institute of Radio Engineers,* and his discovery of a source of radio noise in the stars even made the front page of the *New York Times,* and then, in 1934, he moved on.

In the following decade the number of readers who found his findings provocative enough to pursue was exactly one, another twenty-six-year-old radio engineer, this time in the Chicago area. One day in June 1937 Grote Reber walked out to the yard behind his home in Wheaton, Illinois, and began construction on a tiltable antenna in the shape of a parabola, a surface that, as several centuries of astronomical work with lenses and mirrors had shown, offered the most efficient surface for gathering and focusing signals. The supports were wooden two-by-fours; the dish itself was galvanized iron, thirty-one feet across, and taller than a house. When he completed the towering structure that September, Reber discovered that the spark of nearby automobile ignitions interfered with his observations. So he decided to conduct his research only at night, between the hours of midnight and 6:00 A.M. Six years later he had completed the first radio map of the Milky Way galaxy.

During that time he had become a familiar presence among the astronomers at the Yerkes Observatory, which was, thanks to George Ellery Hale, one of the few places where Reber's work could expect to find a reasonably sympathetic hearing. He eventually convinced the editor of *Astrophysical Journal* (which by now bore the words "Founded in 1895 by George E. Hale and

James E. Keeler" on the cover of every issue) to publish his article "Cosmic Static" in 1940. He followed that article with another article, called "Cosmic Static," and then a third, called "Cosmic Static," and finally a fourth, "Cosmic Static." His persistence paid off. World War II spurred the development of radio technology, and static itself played a role in the war when strange radio signals that the Allies had assumed to be German jamming turned out to be the result of solar flares. In the late 1940s engineers who had worked with radio during the war learned to turn to Reber's writings as they fiddled with this unsettling new technology, trying to develop it, just to hear what they could hear.

Actually engineers for the most part *didn't* listen to radio waves (though Jansky had). It was easier for engineers to translate the signals into tracings of ink on paper. Still, the formal search of the skies that the radio engineers of the postwar era were initiating was as much a voyage into the unknown as any other in the history of astronomy. In Britain an engineer and war veteran named Bernard Lovell persuaded the war-depleted government to sponsor Jodrell Bank, a radio telescope with a diameter of 250 feet, without knowing what he could deliver for that investment. During one of the initial observations at Jodrell Bank, in August 1950, two astronomers aimed the antenna at the M31 spiral galaxy in Andromeda. "The rest of us at Jodrell watched with incredulity as the evidence for radio emission from the nebula began to emerge," Lovell later recalled. "Radio waves had been detected from a galaxy two million light-years distant, and it was no longer possible to regard the local galaxy as in any way unique as a radio emitter."

In effect, Jansky, Reber, and the other pioneers had invented a new form of telescope, the radio receiver. Even as George Ellery Hale's last next, best project, the 200-inch telescope at Mount Palomar (named the Hale Telescope at its dedication on June 3, 1948), was assuming the distinction of largest telescope in the world, a wonder in its own right capable of photographically capturing stars ten million times fainter than those visible to the naked eye, and a worthy successor to all the mighty mirrors of Hale and Herschel, a new and entirely unforeseen form of telescopic instrumentation was giving birth to a new kind of astronomy, indeed another New Astronomy, and yet one more entirely new understanding of the universe.

Astronomers were going to have to learn to see all over again. Since the time of Galileo they had been relying on a magnificent new instrument to deliver what they knew to be nonetheless a small percentage of the potential light. In fact, until the advent of photography, that amount was around 1 percent, and even the most sensitive photographic plates had raised it, at best, to 10 percent. But now astronomers were discovering that what they'd been observing 1 percent *of* was itself only a fraction of the available wavelength information: Visible light covered only about 2 percent of the electromagnetic spectrum. So for more than three centuries astronomers had been studying 1 percent of 2 percent of the potential data waiting out there, and calling that the universe—the cosmic equivalent of judging a book by its cover.

Astronomers were going to have to learn to *think* all over again. The branch of science that had arisen out of a desire to see more—to get closer, to gather more light—was now embarking on a wholly unexpected adventure into the invisible, a direction that was sure to expand the definition of seeing, stretch the concept of the telescope, and strain belief in ways that astronomers couldn't begin to anticipate. Accordingly, *this* New Astronomy was perhaps even more deserving of the encomium than the nineteenth-century version. As one historian of radio astronomy wrote, astronomers at mid-century found themselves entering "an experience comparable in its totality only with Galileo's first sight of the night sky through a telescope."

For hundreds of years astronomers had been playing with light—seeking it, losing it, bouncing it, bending it, timing it, gathering it, scrutinizing it. As of the mid-nineteenth century and the advent of photography, they'd begun accumulating it. Now they were even redefining it, for the time had come that further headway would be difficult unless everyone could first agree on the fundamental source of information, light itself: What was it?

Among astronomers, light actually had started going by another name in the previous century, electromagnetic radiation. In 1800 William Herschel had noticed that when he worked with filters to protect his eyes during his research on the Sun, different colors of glass seemed to produce differet sensations of heat. He tried an experiment. Like Newton in 1666, he passed white light through a prism and found arrayed before him the familiar spectrum of colors, from violet through red;

unlike Newton, he placed a thermometer at various points along the spectrum. As he suspected, the temperatures differed. The lowest were at the violet end of the spectrum and rose toward the red—and then, beyond the end of the spectrum, kept rising. The following year J. W. Ritter used chemical reactions to prove the existence of rays back at the other end of the spectrum, beyond the violet. Not only had Herschel proved what he'd set out to prove (that light and radiant heat are identical, itself a momentous finding), but he and Ritter had discovered infrared and ultraviolet radiation extending beyond either end of the visible spectrum—or, as Herschel wrote, radiant heat that consists, "if I may be permitted the expression, of invisible light."

Visible light, it turned out, was in the minority. The following year, 1802, Thomas Young discovered that to the human eye, the difference between colors is a difference in wavelength—the distance from crest to crest in the undulations of a wave. The wavelengths to which our eyes are sensitive happen to be slightly more and slightly less than 1/20,000 of an inch, growing shorter along the spectrum from red to violet. More than sixty years later the Scottish physicist James Clerk Maxwell interpreted light in terms of electrical and magnetic radiation and suggested that further research might reveal radiation at wavelengths both shorter and longer than visible light. In the coming decades, it did: wavelengths thousands and millions of times shorter—X rays and gamma rays—and millions of times longer—radio waves. And once astronomers discovered practically by accident just how promising radio waves might be as a source of celestial investigation, they had to ask themselves whether the other kinds of invisible light would yield their own kinds of information.

The first problem was how to reach it. Except for radio waves and visible light, the vast majority of electromagnetic radiation from space is blocked by the Earth's atmosphere, leaving astronomers who want to study it little choice but to go up and retrieve it. After World War II, researchers used a cache of twenty-five captured samples of a recent German military innovation, the V-2 rocket, to get their first glimpse of the heavens in several bands of invisible light. Instead of bearing explosives across the English Channel along gentle trajectories that eventually intersected with British soil, these rockets carried photographic emulsions, Geiger counters, and other radiation-sensitive "telescopes" straight up into the skies over the New Mexico desert. In 1946 astronomers found the first evidence of ultraviolet radiation in the Sun. Two years later they confirmed that our local star is also a source of X rays. But it was only with the onset of the space age proper—when rockets reached the velocity necessary to escape the gravitational pull of the Earth—that astronomers could begin to search the universe outside the solar system for sources of invisible light. In 1962 a rocket outfitted with a Geiger counter located the first extrasolar sources of X rays, including one with an output ten *billion* times that of the Sun.

Radio, infrared, optical, ultraviolet, X ray, and gamma ray: What, if anything, did each stop along the spectrum of light have to tell astronomers about the heavens?

When astronomers look at the electromagnetic spectrum, they see the evidence of energy, of heat, that William Herschel had first detected in 1800, and they see this energy or heat varying as they move in one direction or the other along the spectrum: The longer a wavelength, the less energetic, and therefore cooler, it is; the shorter a wavelength, the more energetic, or hotter. In order to make sense of the universe that invisible light was revealing to them, astronomers in the third quarter of the twentieth century realized that in effect they were going to have to figure out which celestial phenomena correspond to what temperatures.

For observations within the Milky Way galaxy, these determinations were relatively straightforward, however improbable the results. At the radio end of the spectrum, home of the longest wavelengths and therefore lowest energy, astronomers in the 1960s found evidence of objects they called pulsars—dead, collapsed stars rotating dozens, even hundreds of times a second. In the infrared, next up the spectrum in terms of wavelength and energy, two researchers at the California Institute of Technology went looking for stars that would be cooler, and therefore older, than stars available only in optical light. Their doubting colleagues warned them they'd be lucky to find even a few dozen; when they'd finished their survey six years later, they'd counted 5,612. Ultraviolet light, on the other side of optical light from the infrared, corresponds to hotter, and therefore younger, stars—newborns, even. X rays, signifying temperatures in the millions of degrees, suggest infernos belonging to exploding stars. Finally, gamma rays, the shortest and therefore most energetic waves on the spectrum, must be remnants of nuclear reactions—collisions between atomic particles—and in fact the first extensive detection of gamma rays came in the 1970s by U.S. spy satellites looking for evidence indicating clandestine Soviet nuclear testing. They found it, all right, but it was coming from the wrong direction—deep space, though how deep astronomers couldn't say. If the gamma-ray sources lay in or near the Milky Way galaxy, then they were powerful indeed. But if they somehow lay beyond the galaxy, in the farthest depths of space and time, they would represent the most powerful sources of energy in the universe—phenomena that could radiate at the speed of light for several billion years and still pack the wallop of a nuclear blast.

In fact, all observations of invisible-light sources *outside* the Milky Way galaxy required adjustments that took into account for their vast distances. In determining the temperature of an extragalactic object millions or billions of light-years away, astronomers had to factor in how much the energy would have cooled, how much the wavelengths would have stretched, before reaching their telescopes. Across such tremendous distances, astronomers realized, even ultra-low-energy radio waves would have begun their journey at high temperatures. In 1958, radio astronomers began picking up faint glimmers from galaxies so distant that, according to their calculations, the radiation must have originated as energy bursts equivalent to millions of Suns, a level of production that confounded their knowledge of physics. By the early 1960s, radio astronomers had found similar readings from *individual* points of light that optical observations always had led them to believe were stars but that now turned out to be extragalactic. Astronomers dubbed them quasi-stellar radio sources, or quasars, an entirely new category of celestial phenomenon. And what to make of extragalactic readings at shorter wavelengths, at even higher levels of energy? Infrared and ultraviolet observations allowed them to visit previously invisible interiors of galaxies while X rays at cosmic depths, astronomers gradually realized, must have originated as the "dying screams" of objects disappearing into black holes—previously hypothetical phenomena so gravitationally voracious that they swallow even light.

Somehow astronomers had gotten the universe wrong—not all wrong, to be sure, but at least enough to render their understanding of it shockingly incomplete. Not only was the universe full of information far more extensive than the naked eye could see even with the help of optical telescopes, but it also was far more active—turbulent, convulsive, violent. Just as astronomers were learning to accommodate the idea that the universe is expanding, they began to observe it in invisible light, and in so doing to fathom the full implication of Edwin Hubble's finding: The universe is a work in progress.

When William Herschel marveled in 1813 that he had observed light that had traveled for two million years— that he had seen two million years into the past—it was the beginning of the end of Newton's clockwork tranquillity. Such idle speculation inevitably raised the question: If that's what the source of light looked like two million years ago, what did it look like now? Did it even exist? In order to find out the answer, however, we would need to be somewhere else—not here on Earth but there, two million light-years distant. In other words *now,* in terms of that star, could only be *there.* In this way astronomers began to comprehend that where we are in space relative to what we're observing *is* where we are in time. It's true when we look at the Sun, only eight minutes from Earth; when we look at Herschel's source of light two million years away; or when we look at a galaxy thousands of millions—that is, billions—of years distant.

Even so, the idea that a universe as old as its most distant objects might be changing over time simply hadn't occurred to anyone until Hubble came along, looked through Hooker, and showed that it was—or if the idea had occurred to anyone, it hadn't seemed credible. Ever since Newton had arrived at a universal law of gravitation, the common assumption had been that while the positions of the individual parts might exhibit relatively small motions over time—a tug here, a pull there, as they followed their predictable elliptical paths—the cosmos had always existed pretty much the way it was now. Newton had supplied his clockwork universe with a sometime stem-winder, a God who rarely, yet occasionally, had to intervene to keep the whole contraption from collapsing on itself. Pierre Simon Laplace, the mathematician who tried to account for every movement in the solar system while working from his study, later triumphantly removed the need for a Creator, proving that Newton should have had more faith in his physics, but he too failed to anticipate the full implications of what he was celebrating. Even Albert Einstein, in figuring out his general theory of relativity, had arrived mathematically at an expanding universe a dozen years before Hubble did so observationally, but Einstein too refused to believe his results. Instead he'd inserted a phantom figure, a fudge factor—what Einstein called a "cosmological constant"—into his equations so that in effect, the universe wouldn't be expanding mathematically after all. When another mathematician argued in the early 1920s that an expanding universe might be unavoidable mathematically, Einstein had acknowledged the possibility but refused to accept it.

Such resistance was understandable. Aristotle's geocentric universe had survived as long as it had because, given the available evidence, it made the most sense, and it became obsolete only when Galileo's observations through the telescope provided new, more compelling evidence. This new heliocentric, Copernican universe as interpreted by Newton then made the most sense, and it would survive unless and until new observations came along to render *it* obsolete.

It might have been forever; it turned out to be 1929. A year after Hubble had released his puzzling preliminary redshift finding that the farther a galaxy is, the faster it's receding, Einstein happened to be in residence at Mount Wilson, where he could monitor the latest results from the 100-inch and discuss them with George Ellery Hale, among others. "Whether the increasing red shift of the lines with distance means actual motion or not," Hale wrote in a letter at this time, "is one of the subjects that greatly interests Einstein, who is ready to accept any explanation that proves to be best supported by all the available evidence." Einstein himself later called his initial resistance to the implications of his own work "the biggest blunder of my life."

A universe that was expanding, however, must be changing *from* something. In the 1950s George Gamow, a Russian-born American working with two other American physicists, Ralph Alpher and Robert Herman, proposed a model of the universe that traced the expansion back by way of mathematics, ending (which was to say, beginning) with an infinitely impacted bundle of everything about the size of a grapefruit. That bundle blew up to become what we are today—but "blew up" *not* in the sense of an explosion. Instead Gamow thought his universe blew up more like a balloon, a curved surface expanding ever outward, each point on the surface separating from every other molecule, so that from the point of view of any one point, or galaxy, every other point would be speeding away from it. It wasn't as if new "space" were being created to fill the gaps between galaxies; instead already existing space was constantly expanding, just as every point on the surface of an inflated balloon was present at first breath. Still, rival astronomer Fred Hoyle derisively called Gamow's model a "big bang" on BBC radio; the term stuck, and so did a widespread misconception that the big bang referred to an exploding (rather than expanding) universe, but not Hoyle's implicit insult—especially after a major prediction of the big bang theory came to pass.

If the big bang model was right, the energy from that initial expansion would still be present. The universe would still be resonating with residual "noise" from all that turbulence—the light that over billions of years had "cooled," or stretched, or redshifted along the electromagnetic spectrum into the far end of the radio waves. Specifically, Alpher and Herman determined that these radio waves should exhibit a wavelength corresponding to a temperature only around 3 degrees above absolute zero, the temperature of −273 degrees Celsius at which all thermal movement of atoms and molecules stops. In 1964 Arno Penzias and Robert Wilson, two radio engineers at (again) the Bell Labs facility in Holmdel, New Jersey, were testing a new, extremely sensitive twenty-foot antenna for use in satellite communications when they detected an odd radiation. After eliminating several explanations for this new mystery hiss (including clearing the equipment of "a white dielectric substance," or pigeon droppings), they asked for help from other engineers in the field. One, Robert Dicke at nearby Princeton University, had independently arrived at the same 3-degree prediction as Alpher and Herman, and he recognized in the Bell Labs data the wavelength and intensity equivalent to this key temperature. Dicke and his group joined forces with Penzias and Wilson and confirmed that they'd found the "three degree background."

This detection did more than validate the big bang theory. For the second time in the same century at the

same research facility, an extraordinarily important telescopic discovery had taken place quite by accident by
engineers using the *same new technology.* Astronomers
had to wonder what they might be able to find with a little more planning. Specifically, now that they had begun
to figure out what each new source of information could
tell them, they had to figure out how to get it to tell them
more.

For not only had the detection of the "three degree
background" further validated invisible-light astronomy
but it had suggested its potential as an empirical science—one that would need new forms of instrumentation, triumphs of engineering that resembled traditional
telescopes only in that the primary purpose remained the
relentless accumulation of more light. The extremely
long waves at the radio end of the spectrum required
dishes with diameters far beyond any sizes that optical
astronomers had ever foreseen. The Arecibo radio telescope in Puerto Rico, for instance, spanned a thousand
feet across, giving it a surface area large enough to contain twenty-six football fields. Infrared radiation presented an entirely different problem. It proved to be so
prevalent (human beings glow with it, as do the inani-
mate objects around them) that detectors could block
their own radiation only by being maintained at a temperature near absolute zero. And astronomers hoping to
capture X rays, which would shred the mirror of an optical telescope, learned to utilize tubular mirrors that
deflect, not reflect, the radiation.

At the same time, the invisible-light revolution forced
a thorough rethinking of the phenomena available in

visible light, and astronomers found themselves wishing for instruments that would allow them to see seemingly familiar objects in new detail and to see new objects altogether—that is, for instruments with higher resolutions and greater capacities for gathering light. In these endeavors they were aided by twentieth-century telescopy's answer to the micrometer and the telescopic sight—the computer. Beginning in the 1970s, the glass photographic plate of the nineteenth century became virtually extinct, to be replaced by the charge-coupled device, or CCD, an extremely light-sensitive instrument that not only stored information digitally but was also five times more sensitive than previous electronic optical devices, which themselves were ten times more sensitive than photographs alone. Rather than 2 percent or 10 percent, some observatories in a matter of years became capable of 50 percent to 80 percent efficiencies—while still using the same old telescopes.

Astronomers also figured out new ways to widen apertures: a honeycomb design in Arizona full of mirrors that were three-quarters air, or, in Hawaii, the twin 10-meter Keck Telescopes, which were comprised of 36 hexagonal segments, each 1.8 meters across, that a computer corrected hundreds of times a second to simulate two perfect surfaces. Computers allowed astronomers to manipulate light before they'd gathered it, too. Adaptive optics recorded changes in the Earth's atmosphere and adjusted the primary focus automatically. Active optics monitored a mirror deforming under its own weight and adjusted the telescope's supports instantaneously. Fiber optics followed an image of several thousand galaxies and adjusted the focus on a nebula-by-nebula basis.

Computers also allowed astronomers to continue to alter the image after the telescope had retrieved it—or, more accurately, the digital data corresponding to an image. They could highlight temperature variations by color, or eliminate the light from a distracting galaxy in the foreground to reconstruct a clear image of a galaxy in the background. Computers could also construct cosmological models, taking physicists back to a fraction of a fraction of a second after the big bang, or one hundred trillion years into the future.

In professional circles at least, not only had the photographic plate long ago displaced the astronomer at the eyepiece, but on certain occasions the computer chip now allowed astronomers to remove themselves from the observatory altogether. Instead of braving the bracing night air on a viewing platform a thousand miles from home, astronomers could sit at a terminal in their own offices, manipulating the telescope from their keyboards, or simply showing up in the morning and downloading data.

And what data: visible-light detection of two new moons of Uranus; radio wave evidence of planets in other solar systems, detectable by their gravitational effects on their host stars; X-ray evidence of black holes; X-ray evidence of black holes at the center of our galaxy, and perhaps every galaxy; X-ray, infrared, and radio evidence of a black hole in action, alternately devouring everything nearby and spouting geysers of gas at 90 percent the speed of light; X-ray evidence of an object dragging space and time around itself, confirming that particular prediction of Einstein's general theory of relativity; radio-wave evidence of a "nest" of at least a dozen super supernova explosions, each expending hundreds of times more energy in one brilliant burst than the Sun will in its entire lifetime; gamma ray, X-ray and optical evidence that gamma ray bursts do indeed originate in the farthest recesses of space and are therefore the most powerful known sources of energy in the universe; a radio wave picture of Penzias and Wilson's "three degree background" showing the ripples from the big bang that one day would condense into galaxies, stars, and us—and this was just for starters. The wealth of information itself became a kind of joke: HO HUM, MORE NEW PLANETS, read the headline in one astronomy magazine.

It's always tempting for each generation to imagine itself at the heart of a golden age, almost as if the inability to be at the center of the universe in spatial terms fosters a need to be at the center in temporal terms. But the specific reason for the outpouring of evidence in the 1980s and 1990s in fact did suggest a parallel to the last golden age.

In many respects the telescope was right back where it began, in the first decades of the seventeenth century. First had come a relatively crude qualitative phase. What the mountains on the Moon, the Medicean planets, and the phases of Venus meant to Galileo, so pulsars, black holes, and gamma ray bursts signified for the astronomers of the 1950s through the 1970s. Next had come the more delicate quantitative phase, and just as the astronomers of the last half of the seventeenth century determined the distances and dimensions of the solar system, so the astronomers at the end of the twentieth century had begun to take the measure of their universe.

The Hipparcos satellite survey, begun in 1989 and completed in 1997, determined the once-elusive parallax for 118,218 stars, and provided the position of 1,058,332 altogether, triple the number in any previous atlas of the Milky Way galaxy. "Redshift surveys" measured distances to hundreds of thousands, then hundreds of millions, of individual galaxies, not only endowing the universe with the third dimension that Herschel had granted our galaxy, but discovering that galaxies themselves belong to clusters, that these clusters in turn belong to superclusters nearly a billion light-years across, and that these superclusters seem to exhibit even larger structures—walls or filaments that suggest the universe itself might well resemble nothing so much as a molecule. And a year after the Hubble Space Telescope drilled a hole through the heavens and astronomers calculated the number of galaxies at fifty billion, they took a closer look at their data and calmly doubled the estimate.

But what truly distinguished the outpouring of quantitative data at the close of the twentieth century wasn't just the census information, and it wasn't even the third dimension that it suggested at the farthest reaches of the universe, but it was the reconsideration of the cosmos in a *further* dimension. In its distances, temperatures, and velocities—in the force of its furies, in its births and deaths, indeed in its birth and death—the universe came alive in time.

It was some measure of just how new this model was that the answer to the question of where was the end of the universe was no longer *there,* the stellar vault, or even *there,* beyond the galaxy, but *then:* the beginning of time. Two years after the release of the Hubble Deep Field, an astronomer announced that he had been able to detect nothing between the galaxies—that the telescope had indeed reached the limits of visible light. As for *invisible* light, another astronomer at the same meeting reported that an infrared survey of the sky, conducted aboard a satellite for ten months in 1989 and then ana-lyzed for eight years, had measured the total amount of energy released since the beginning of time by all the stars in the universe (in all galaxies, not just our own). The result was double what even the Hubble Space Telescope had found. Perhaps the rest was hiding behind dust, which the infrared telescope aboard the satellite would have been able to see through, or was faint enough or far enough away to escape detection. Whatever, the result served as a useful reminder. As the astronomer said, "When the Hubble Deep Field came in, we got into the mindset that we were seeing it all."

Even so, nobody was claiming that *this* was it, then: journey's end. In the late 1970s, astronomers using both optical and radio observations found galaxies spinning in such a way that they seemed to be violating Kepler's law regarding velocity of rotation: Their outer rims were simply going too fast. The only way that what they were seeing could make sense gravitationally was if the disks they were observing were larger than they appeared, if some other invisible matter were there, on the outskirts of the rotation, accounting for the unusual speeds of the visible inner ring. This invisible matter, however, did not appear—not in the optical, or in the radio, or in any other part of the electromagnetic spectrum.

This was a mystery to rival Jansky's hiss, if not surpass it: a missing mass that operated outside all wavelengths. It turned out not to be just invisible; it was undetectable, except through the gravitational effect it had on matter that astronomers *could* detect. Scientists christened this missing majority of the universe "dark matter" and estimated that it constituted 90 percent of the universe, maybe 99 percent.

So for most of the history of the telescope, astronomers had been studying 1 percent of 2 percent of, at most, 10 percent of what's out there, and calling *that* the universe.

In 1936 Edwin Hubble wrote, "The history of astronomy is a history of receding horizons." He was referring to the depths of space that astronomers had sequentially assumed to be the extent of their investigations: first what he called "the realm of the planets," then "the realm of the stars," and eventually, thanks to his own efforts, "the realm of the nebulae." In fact, once Hubble plotted the redshifts of a handful of nebulae, the idea of a static and unchanging universe came to seem as primitive in its own way as the Aristotelian dome of fixed and unchanging stars. It had survived as long as it had only because it seemed to make the most sense. Now,

Hubble had rendered it obsolete by providing a new set of observations to support a new model of the universe. It was a model that was violent and always changing; that still appeared to follow a few simple rules, but along spacetime paths and down black holes that Pierre Simon Laplace never could have imagined; that turned out not to be geocentric or heliocentric but noncentric—or, perhaps, omnicentric: Since everything was present in the big bang, everything is technically at the center of the universe. Or nothing is. Either way, once Hubble saw galaxies racing away from him in every direction, this was the model that made the most sense, that offered the best match for what Galileo once called "the certainty of sense evidence."

The flatness of Earth, the immobility of the globe, the perfectly circular paths of the planets, the Earth's position at the center of the Kosmos, the Sun's position at the center of a new cosmos, the fixed positions of the stars, the stars themselves as the extent of the universe, the universe as all that we could ever hope to see, and all that we could ever hope to see as all there is—all the assumptions safely based on observation were gone. Had we committed a blunder as fundamental in our conception of the universe as the one that mapmakers had committed several centuries earlier, defining the world, from their warm Mediterranean shores, strictly in terms of those lands they could visit?

The technology of the telescope had always set limits on the available information, and observers had always used those limits to reinforce prevailing beliefs even while advancing new ones. The limitations of the

Galilean telescope confirmed what seventeenth-century observers knew: that Galileo had seen everything worth seeing. Similarly, the limitations of the Keplerian telescope confirmed what eighteenth-century astronomers knew: that the stars held no surprises. Now, in the aftermath of the New Astronomy of invisible light, to say nothing of the dark matter revolution, it turned out that the instrument in any incarnation came with its own inherent set of limitations, not on what to expect or where to look, but on *how* to look—not on the available information, but on the nature of the information itself.

The promise of technology was a false promise, but it was not an empty promise. The conclusions it fostered weren't necessarily wrong, only incomplete. In its first four centuries the telescope helped create a science that took our view of the universe from one that, in current terminology, was no bigger than the solar system, to one that was no bigger than the galaxy, to one that was at least as big as superclusters of galaxies—a hundred billion, to be approximately exact. And if we never reached the ends of the universe, if we never divined what came before the big bang or what lurks beyond the observable horizon, we at least learned to put our inability to do so in a new perspective. Whenever we couldn't conceive of what's out there, whenever we couldn't even begin to guess, it wasn't only because we still lacked the technology, and it wasn't only because we still lacked the information, but it was because we didn't yet understand what the preconceptions might be that were restricting our view.

When Galileo first pointed his tube of long seeing at

the Moon, in a garden of Padua on a clear autumn evening in 1609, the question of whether the heavens were variable might have been present, but not pressing. Instead he learned what questions to ask by looking at the answers. It wasn't long before the telescope itself seemed as if it might be an answer of some sort, the solution to a long-forgotten riddle about seeing distant objects as if they were near, or, after it had developed into an instrument capable of exquisite precision measurements, a vessel for delivering on a promise of perfect understanding. But after nearly four hundred years it had come to seem like a method as well, a means toward an end—at least, an end for now: a way to begin to explore the assumptions that confine us, the givens that give us away, the Ocean River that forever encircles us.

The answer, it turned out, was what we don't know, the question was the telescope, and the rest was history.

Bibliography

It's no coincidence, I suspect, that the study of the history of science is a twentieth-century, and particularly a post-Einstein, post-Hubble, phenomenon. Only after the collapse of Newton's cosmos, only after the end of a belief in the ultimate acquisition of knowledge, could the question of "What's next?" become pressing—and in order to formulate an answer it became necessary to figure out what had come before. For three centuries New Philosophers and their intellectual descendants had labored under the assumption that they were completing the job started by ancient scholars, that if only they applied themselves hard enough and long enough they would relieve the universe of every last one of its mysteries. Then came relativity and the expanding universe (to say nothing of quantum mechanics and the uncertainty principle) to disabuse them of their unquestioning faith and to create the need to do with the previous three or four centuries what their Renaissance predecessors had done with their own histories: put the past in perspective.

So blindingly persuasive were the results that they've made a preperspective view of the past four centuries frustratingly remote, if not forever inaccessible. For this reason, readers might be as surprised as I was to learn that as recently as the 1920s the journal *Nature* was running letters debating the appropriateness of calling a person who practices science a "scientist." The first reference to the "Scientific Revolution" dates from the 1930s, and the first doctorate in the history of science from 1943. It was the members of this first generation of "historians of science" (whether academically certified as such or not) who produced the works through which we can't help seeing and interpreting much of the past millennium, and while these classics have inevitably inspired further revisionism, or at least reconsiderations (sometimes by their own authors), they're still a good place to start. They were for me, anyway: Marie Boas's *The Scientific Renaissance 1450–1630*, Herbert Butterfield's *The Origins of Modern Science 1300–1800*, A. Rupert Hall's *The Scientific Revolution, 1500–1800*, Alexandre Koyré's *From the Closed World to the Infinite Universe*, and Thomas S. Kuhn's *The Structure of Scientific Revolutions*.

This bibliography is comprehensive regarding only my research. Even so it includes many of the standard works, or at least works by the leading experts, that a reader might find useful in pursuing one of the major historical figures in this book. For Galileo, see Stillman Drake; for William Herschel, Michael Hoskin; for George Ellery Hale, Helen Wright. In a more general sense, the two books by Michael J. Crowe provide excellent introductions to, as well as lengthy selections from, the astronomical writings that most influenced man's conception of the universe.

This book attempts a general overview of the history of the telescope with an emphasis on its cosmological implications. Readers seeking a more technical approach to the telescope should consult the Louis Bell. A more popular rendering is available in either the Isaac Asimov or Richard Learner. The Henry C. King, though hardly current, is still the definitive history of the instrument.

Anyone seriously interested in the philosophical implications of this instrument, however, sooner or later will encounter Albert Van Helden, whose prodigious body of scholarly work on the subject deserves a mention all its own.

For reasons of space and sanity, I have excluded from this bibliography (with the occasional essential exception) magazine and newspaper articles. This is not to slight *Astronomy, Discover, Natural History, Science News, Scientific American, Sky & Telescope,* or any other publication, printed or online, that on the whole appeals more to a general than a scholarly readership. In fact, these publications were invaluable resources, as they would be for any reader wishing to explore the subject of astronomy.

Abetti, Giorgio. *The History of Astronomy.* New York: Henry Schuman, Inc., 1952.

Armitage, Angus. *Sun, Stand Thou Still.* New York: Henry Schuman, Inc., 1947.

————. *William Herschel*. Garden City, N.Y.: Doubleday & Company, Inc., 1963.

Asimov, Isaac. *Eyes on the Universe*. Boston: Houghton Mifflin Company, 1975.

Bell, Louis. *The Telescope*. New York: Dover Publications, Inc., 1981 (reprinted from 1922).

Bennett, J. A. " 'On the Power of Penetrating into Space': The Telescopes of William Herschel," *Journal for the History of Astronomy* (June 1976), pp. 75–108.

Berendzen, Richard, Richard Hart, and Daniel Seeley. *Man Discovers the Galaxies*. New York: Science History Publications, 1976.

Bernstein, Jeremy. *A Theory for Everything*. New York: Copernicus, 1996.

Bloom, Terrie F. "Borrowed Perceptions: Harriot's Maps of the Moon," *Journal for the History of Astronomy* (June 1978), pp. 117–22.

Boas, Marie. *The Scientific Renaissance 1450–1630*. New York: Harper & Brothers, 1962.

Boorstin, Daniel J. *The Discoverers*. New York: Vintage Books, 1985.

Brinton, Crane, ed. *The Portable Age of Reason Reader*. New York: Viking Press, 1956.

Bronowski, Jacob. "Copernicus as a Humanist," in *The Nature of Scientific Discovery,* ed. Owen Gingerich. Washington: Smithsonian Institution Press, 1975.

Burke, James. *The Day the Universe Changed*. Boston: Little, Brown and Company, 1985.

Burke, John G., ed. *The Uses of Science in the Age of Newton*. Berkeley: University of California Press, 1983.

Butterfield, Herbert. *The Origins of Modern Science 1300–1800*. London: G. Bell and Sons Ltd., 1958.

Christianson, Gale E. *This Wild Abyss*. New York: Free Press, 1979.

————. *Edwin Hubble*. New York: Farrar, Straus, Giroux, 1995.

Coffin, Charles Monroe. *John Donne and the New Philosophy*. New York: Columbia University Press, 1937.

Cohen, I. B. "Roemer and the First Determination of the Velocity of Light (1676)," *Isis* (April 1940), pp. 327–79.

Cohen, I. Bernard. *Science and the Founding Fathers.* New York: W. W. Norton & Company, 1995.

Copernicus, Nicholas. *Three Copernican Treatises,* translated with introduction and notes by Edward Rosen. New York: Dover Publications, Inc., 1959.

Crowe, Michael J. *Theories of the World from Antiquity to the Copernican Revolution.* New York: Dover Publications, Inc., 1990.

————. *Modern Theories of the Universe, from Herschel to Hubble.* New York: Dover Publications, Inc., 1994.

Dampier, Sir William Cecil. *A Shorter History of Science.* Cleveland: World Publishing Company, 1957 (reprinted from 1944).

Drake, Stillman. *Galileo Studies.* Ann Arbor: University of Michigan Press, 1970.

————. "Galileo's First Telescopic Observations," *Journal for the History of Astronomy* (October 1976), pp. 153–68.

————. *Galileo at Work: His Scientific Biography.* Chicago: University of Chicago Press, 1978.

Dressler, Alan. *Voyage to the Great Attractor.* New York: Vintage Books, 1995.

Dreyer, J. L. E. *A History of Astronomy from Thales to Kepler.* New York: Dover Publications, Inc., 1953 (reprinted from 1906).

Durant, Will. *The Story of Philosophy.* New York: Simon & Schuster, 1961.

Eddington, Sir Arthur. "Weighing Light," in *Astronomy,* eds. Samuel Rapport and Helen Wright. New York: New York University Press, 1964.

Einstein, Albert. *Relativity: The Special and the General Theory,* translated by Robert W. Lawson. New York: Crown Publishers, Inc., 1961.

Elliott, J. H. *The Old World and the New 1492–1650.* Cambridge: Cambridge University Press, 1970.

Fahie, J. J. *Galileo—His Life and Work.* London: John Murray, 1903.

Fehrenbach, Charles. "Twentieth-Century Instrumentation," in *Astrophysics and Twentieth-Century Astronomy to 1950,* Part A, ed. Owen Gingerich. Cambridge: Cambridge University Press, 1984.

Ferris, Timothy. *Coming of Age in the Milky Way.* New York: William Morrow and Company, 1988.

———. *The Whole Shebang.* New York: Simon & Schuster, 1997.

Field, J. V., and Frank A. L. J. James, eds. *Renaissance and Revolution.* Cambridge: Cambridge University Press, 1993.

Forbes, Eric G. "Early Astronomical Researches of John Flamsteed," *Journal for the History of Astronomy* (June 1976), pp. 124–38.

Friedman, Herbert. *The Astronomer's Universe.* New York: Ballantine Books, 1990.

Galilei, Galileo. *Discoveries and Opinions of Galileo,* translated with an introduction and notes by Stillman Drake. New York: Anchor Books, 1957.

———. *Dialogue Concerning the Two Chief World Systems—Ptolemaic & Copernican,* translated by Stillman Drake, foreword by Albert Einstein. Berkeley: University of California Press, 1967 (reprinted from 1953).

———. *Sidereus Nuncius, or The Sidereal Messenger,* translated with introduction, conclusion, and notes by Albert Van Helden. Chicago: University of Chicago Press, 1989.

Gillispie, Charles Coulston, ed. *Dictionary of Scientific Biography.* New York: Charles Scribner's Sons, 1970–1990.

Gingerich, Owen. *"Dissertatio cum Professore Righini et Sidereo Nuncio,"* in *Reason, Experiment and Mysticism,* eds. M. L. Righini Bonelli and William R. Shea. New York: Science History Publications, 1975.

———, ed. *The Nature of Scientific Discovery.* Washington: Smithsonian Institution Press, 1975.

———, ed. *Astrophysics and Twentieth-Century Astronomy to 1950,* Part A. Cambridge: Cambridge University Press, 1984.

———. "A Copernican Perspective," in *The Nature of Scientific Discovery,* ed. Owen Gingerich. Washington: Smithsonian Institution Press, 1975.

———. "Does Science Have a Future?" in *The Nature of*

Scientific Discovery, ed. Owen Gingerich. Washington: Smithsonian Institution Press, 1975.

Goldstein, Thomas. *Dawn of Modern Science.* Boston: Houghton Mifflin Company, 1980.

Greene, John C. *American Science in the Age of Jefferson.* Ames: Iowa State University Press, 1984.

Gribbin, John. *In the Beginning.* Boston: Little, Brown and Company, 1993.

Hale, George Ellery. *Beyond the Milky Way.* New York: Charles Scribner's Sons, 1926.

Hall, A. Rupert. *The Scientific Revolution, 1500–1800.* Boston: Beacon Press, 1966.

———. "The Nature of Scientific Discovery," in *The Nature of Scientific Discovery,* ed. Owen Gingerich. Washington: Smithsonian Institution Press, 1975.

Hall, Marie Boas. "The Spirit of Innovation in the Sixteenth Century," in *The Nature of Scientific Discovery,* ed. Owen Gingerich. Washington: Smithsonian Institution Press, 1975.

Halley, Edmond. *Correspondence and Papers of Edmond Halley,* ed. Eugene Fairfield MacPike. New York: Arno Press, 1975 (reprinted from 1932).

Hartner, Willy. "The Role of Observations in Ancient and Medieval Astronomy," *Journal for the History of Astronomy* (February 1977), pp. 1–11.

Hathaway, Nancy. *The Friendly Guide to the Universe.* New York: Viking, 1994.

Hawking, Stephen W. *A Brief History of Time.* New York: Bantam Books, 1988.

Henbest, Nigel, and Michael Marten. *The New Astronomy.* Cambridge: Cambridge University Press, 1996.

Hindle, Brooke. *The Pursuit of Science in Revolutionary America 1735–1789.* Chapel Hill: University of North Carolina Press, 1956.

Holden, Edward S. *Sir William Herschel: His Life and Works.* New York: Charles Scribner's Sons, 1881.

Hoskin, Michael. "William Herschel's Early Investigations of Nebulae: A Reassessment," *Journal for the History of Astronomy* (October 1979), pp. 165–76.

———. *Stellar Astronomy*. Bucks, England: Science History Publications, 1982.

———. "William Herschel and the Making of Modern Astronomy," *Scientific American* (February 1986), pp. 106–112.

Hoskin, Michael A. *William Herschel and the Construction of the Heavens*. New York: W. W. Norton & Company, Inc., 1964.

Jaki, Stanley L. *The Milky Way*. New York: Science History Publications, 1972.

Kelsey, Larry, and Darrel Hoff. *Recent Revolutions in Astronomy*. New York: Franklin Watts, 1987.

King, Henry C. *The History of the Telescope*. New York: Dover Publications, Inc., 1979 (reprinted from 1955).

Koestler, Arthur. *The Sleepwalkers*. New York: Macmillan Company, 1959.

Kolb, Rocky. *Blind Watchers of the Sky*. New York: Addison-Wesley, 1996.

Koyré, Alexandre. *From the Closed World to the Infinite Universe*. Baltimore: Johns Hopkins Press, 1957.

———. *Newtonian Studies*. Cambridge, Mass.: Harvard University Press, 1965.

———. *Metaphysics and Measurement*. Cambridge, Mass.: Harvard University Press, 1968.

———. *The Astronomical Revolution*. New York: Dover Publications, Inc., 1992 (reprinted from 1973 and 1961).

Kuhn, Thomas S. *The Structure of Scientific Revolutions*. Chicago: University of Chicago Press, 1970.

———. *The Copernican Revolution*. Cambridge, Mass.: Harvard University Press, 1979 (reprinted from 1959).

Lankford, John. "The Impact of Photography on Astronomy," in *Astrophysics and Twentieth-Century Astronomy to 1950,* Part A, ed. Owen Gingerich. Cambridge: Cambridge University Press, 1984.

Learner, Richard. *Astronomy Through the Telescope*. New York: Van Nostrand Reinhold Company, 1981.

Lewis, C. S. *The Discarded Image*. Cambridge: Cambridge University Press, 1964.

Lightman, Alan. *Ancient Light*. Cambridge, Mass.: Harvard University Press, 1991.

————. *Time for the Stars*. New York: Viking, 1992.

Lovell, A. C. B. "Radio Telescopes," in *Astronomy*, eds. Samuel Rapport and Helen Wright. New York: New York University Press, 1964.

Lovell, Bernard. *Man's Relation to the Universe*. San Francisco: W. H. Freeman and Company, 1975.

————. *In the Center of Immensities*. New York: Harper & Row, 1978.

————. *Astronomer by Chance*. New York: Basic Books, Inc., 1990.

Lubbock, Constance A. *The Herschel Chronicle*. Cambridge: Cambridge University Press, 1933.

MacPike, Eugene Fairfield, ed. *Hevelius, Flamsteed and Halley*. London: Taylor and Francis, Ltd., 1937.

————. *Correspondence and Papers of Edmond Halley*. New York: Arno Press, 1975 (reprinted from 1932).

Manly, Peter L. *Unusual Telescopes*. Cambridge: Cambridge University Press, 1991.

Meadows, A. J. "The New Astronomy," *Astrophysics and Twentieth-Century Astronomy to 1950*, Part A, ed. Owen Gingerich. Cambridge: Cambridge University Press, 1984.

————. "The Origins of Astrophysics," *Astrophysics and Twentieth-Century Astronomy to 1950*, Part A, ed. Owen Gingerich. Cambridge: Cambridge University Press, 1984.

Meeus, Jean. "Galileo's First Records of Jupiter's Satellites," *Sky & Telescope* (February 1964), pp. 105–06.

Munitz, Milton K., ed. *Theories of the Universe from Babylonian Myth to Modern Science*. Glencoe, Ill.: Free Press, 1957.

Nef, John U. "The Interplay of Literature, Art, and Science in the Time of Copernicus," in *The Nature of Scientific Discovery*, ed. Owen Gingerich. Washington: Smithsonian Institution Press, 1975.

Nicolson, Marjorie. *Voyages to the Moon*. New York: Macmillan Company, 1948.

————. *The Breaking of the Circle*. New York: Columbia University Press, 1962.

————. *Science and Imagination*. Hamden, Conn.: Archon Books, 1976 (reprinted from 1956).

North, John. *Astronomy and Cosmology.* New York: W. W. Norton & Company, 1994.

Oberman, Heiko A. "Reformation and Revolution: Copernicus' Discovery in an Era of Change," in *The Nature of Scientific Discovery,* ed. Owen Gingerich. Washington: Smithsonian Institution Press, 1975.

Ornstein, Martha. *The Role of Scientific Societies in the Seventeenth Century.* New York: Arno Press, 1975 (reprinted from 1928).

Overbye, Dennis. *Lonely Hearts of the Cosmos.* New York: HarperCollins Publishers, 1991.

Pannekoek, A. *A History of Astronomy.* London: George Allen & Unwin Ltd., 1961.

Preston, Richard. *First Light.* New York: Random House (revised edition), 1996.

Rapport, Samuel, and Helen Wright, eds. *Astronomy.* New York: New York University Press, 1964.

Reston, James, Jr. *Galileo: A Life.* New York: HarperCollins, 1994.

Ridpath, Ian. *A Dictionary of Astronomy.* Oxford: Oxford University Press, 1997.

Righini Bonelli, M. L., and William R. Shea, eds. *Reason, Experiment and Mysticism in the Scientific Revolution.* New York: Science History Publications, 1975.

Righini, Guglielmo. "New Light on Galileo's Lunar Observations," in *Reason, Experiment and Mysticism in the Scientific Revolution,* eds. M. L. Righini Bonelli and William R. Shea. New York: Science History Publications, 1975.

Ronan, Colin A. "Galileo Galilei—1564–1642," *Sky & Telescope* (February 1964), pp. 72–78.

———. *Astronomers Royal.* Garden City, N.Y.: Doubleday & Company, Inc., 1969.

———. *Edmond Halley: Genius in Eclipse.* Garden City, N.Y.: Doubleday & Company, Inc., 1969.

Rosen, Edward. *The Naming of the Telescope.* New York: Henry Schuman, 1947.

———. "When Did Galileo Make His First Telescope?" *Centaurus* (1951), pp. 44–51.

————. "The Authenticity of Galileo's Letter to Landucci," *Modern Language Quarterly* (December 1951), pp. 473–86.

————. "Did Galileo Claim He Invented the Telescope?" *Proceedings of the American Philosophical Society* (October 15, 1954), pp. 304–12.

Rowan-Robinson, Michael. *Cosmic Landscape.* Oxford: Oxford University Press, 1979.

Sandage, Allan. "Edwin Hubble 1889–1953," *Journal of the Royal Astronomical Society of Canada* (December 1989).

Schaffer, Simon. "Herschel in Bedlam: Natural History and Stellar Astronomy," *British Journal for the History of Science* (November 1980), pp. 211–39.

Segre, Michael. *In the Wake of Galileo.* New Brunswick, N.J.: Rutgers University Press, 1991.

Shapin, Steven. *The Scientific Revolution.* Chicago: University of Chicago Press, 1996.

Shapley, Harlow. *Flights from Chaos.* New York: Whittlesey House, McGraw-Hill Book Company, Inc., 1930.

————. *Of Stars and Men.* Boston: Beacon Press, 1958.

Shea, William R. "Introduction: Trends in the Interpretation of Seventeenth Century Science," in *Reason, Experiment and Mysticism in the Scientific Revolution,* eds. M. L. Righini Bonelli and William R. Shea. New York: Science History Publications, 1975.

Sheehan, William. *Planets & Perception.* Tucson: University of Arizona Press, 1988.

Sidgwick, J. B. *William Herschel.* London: Faber and Faber Limited, 1953.

Singer, Charles, E. J. Holmyard, A. R. Hall, and Trevor I. Williams, eds. *A History of Technology, vol. III, From the Renaissance to the Industrial Revolution c1500–c1750.* London: Oxford University Press, 1957.

Singer, Charles. *A Short History of Scientific Ideas to 1900.* New York and London: Oxford University Press, 1959.

Smith, R. W. "The Origins of the Velocity-Distance Relation," *Journal for the History of Astronomy* (October 1979), pp. 133–65.

Spangenburg, Ray, and Diane K. Moser. *On the Shoulders of*

Giants: The History of Science in the Eighteenth Century.
New York: Facts on File, 1993.

Sullivan, Woodruff T., III. "Early Radio Astronomy," in *Astrophysics and Twentieth-Century Astronomy to 1950,* Part A, ed. Owen Gingerich. Cambridge: Cambridge University Press, 1984.

Suter, Rufus. "Some Relics of Galileo in Florence," in *Scientific Monthly* (October 1951), pp. 229–34.

———. "Galileo in Padua," *Sky & Telescope* (February 1964), pp. 99–100.

Temkin, Owesi. "Science and Society in the Age of Copernicus," in *The Nature of Scientific Discovery,* ed. Owen Gingerich. Washington: Smithsonian Institution Press, 1975.

Trefil, James. *The Dark Side of the Universe.* New York: Charles Scribner's Sons, 1988.

Tucker, Wallace, and Karen Tucker. *The Cosmic Inquirers.* Cambridge, Mass.: Harvard University Press, 1986.

Turner, A. J. "Some Comments by Caroline Herschel on the Use of the 40ft Telescope," *Journal for the History of Astronomy* (October 1977), pp. 196–98.

Van Helden, Albert. "The Telescope in the Seventeenth Century," *Isis* (1974), pp. 38–58.

———. "The Importance of the Transit of Mercury of 1631," *Journal for the History of Astronomy* (February 1976), pp. 1–10.

———. "The Development of Compound Eyepieces, 1640–1670," *Journal for the History of Astronomy* (February 1977), pp. 26–37.

———. *The Invention of the Telescope.* Philadelphia: American Philosophical Society, 1977.

———. "The Birth of the Modern Scientific Instrument, 1550–1700," in *The Uses of Science in the Age of Newton,* ed. John G. Burke. Berkeley: University of California Press, 1983.

———. "Telescope Building, 1850–1900," in *Astrophysics and Twentieth-Century Astronomy to 1950,* Part A, ed. Owen Gingerich. Cambridge: Cambridge University Press, 1984.

————. "Building Large Telescopes, 1900–1950," in *Astrophysics and Twentieth-Century Astronomy to 1950,* Part A, ed. Owen Gingerich. Cambridge: Cambridge University Press, 1984.

————. *Measuring the Universe.* Chicago: University of Chicago Press, 1985.

————. "Telescopes and Authority from Galileo to Cassini," *Osiris* (1994), pp. 9–29.

————, and Thomas L. Hankins, "Instruments in the History of Science," *Osiris* (1994), pp. 1–6.

Vaucouleurs, Gérard de. *Discovery of the Universe.* New York: Macmillan Company, 1957.

Wallace, Alfred R. *Man's Place in the Universe.* New York: McClure, Phillips & Co., 1903.

Warner, Brian. "Portrait of a 40-foot Giant," *Sky & Telescope* (March 1986), pp. 253–54.

Wheeler, John Archibald. "The Universe as Home for Man," in *The Nature of Scientific Discovery,* ed. Owen Gingerich. Washington: Smithsonian Institution Press, 1975.

Whitaker, Ewan A. "Galileo's Lunar Observations and the Dating of the Composition of 'Sidereus Nuncius,' " *Journal for the History of Astronomy* (October 1978), pp. 155–69.

Wilford, John Noble. *The Mapmakers.* New York: Vintage Books, 1982.

Winkler, Mary G., and Albert van Helden. "Representing the Heavens: Galileo and Visual Astronomy," *Isis* (1992), pp. 195–217.

————. "Hevelius and the Visual Language of Astronomy," in *Renaissance and Revolution,* eds. J. V. Field and Frank A. J. L. James. Cambridge: Cambridge University Press, 1993.

Wolf, A. *A History of Science, Technology, & Philosophy in the 18th Century.* New York: Harper & Brothers, 1961 (reprinted from 1938 and 1952).

Wright, Helen. *Palomar.* New York: Macmillan Company, 1952.

————. *Explorer of the Universe.* New York: E. P. Dutton & Co., Inc., 1966.

————, Joan Warnow, and Charles Weiner, eds. *The Legacy of George Ellery Hale*. Cambridge, Mass.: MIT Press, 1972.

Zajonc, Arthur. *Catching the Light*. New York: Oxford University Press, 1995.

Index